Machine Learning in Air Force Human Resource Management

Volume 2, A Framework for Vetting Use Cases with Example Applications

MATTHEW WALSH, SEAN ROBSON, ALBERT A. ROBBERT,
DAVID SCHULKER

Prepared for the Department of the Air Force
Approved for public release; distribution unlimited.

For more information on this publication, visit **www.rand.org/t/RRA1745-2**.

Published by the RAND Corporation, Santa Monica, Calif.

Library of Congress Cataloging-in-Publication Data is available for this publication.

ISBN: 978-1-9774-1257-7

Cover: Golden Sikorka/Getty Images.

About This Report

The Department of the Air Force (DAF) has begun to develop and field artificial intelligence (AI) and machine learning (ML) systems for myriad mission areas and support functions, including human resource management (HRM). ML systems have the potential to accelerate existing decision processes and to enhance decision quality by leveraging data. Further, by allowing the DAF to make decisions at greater speed and scale, ML systems have the potential to enable entirely new decision processes.

To harness this transformative potential, the DAF must deliberately decide how to allocate limited resources to build a portfolio of AI projects to meet HRM needs. This report describes a framework for evaluating AI use cases for HRM. The framework entails (1) formulating business problems and proposed solutions, (2) characterizing business value or impact, (3) screening out projects that are technically infeasible, (4) assessing the complexity of remaining project proposals, and (5) aligning project proposals with available resources. By using a systematic process, the DAF can better ensure that potential AI projects meet a known HRM need and are likely to succeed. We demonstrate the framework using 19 examples of AI systems aligned with all phases of HRM.

To assist with ongoing efforts to explore ways to use data technology to improve HRM, this project is intended to develop decision-support methods and tools to help managers and panel members process and understand performance records. Given that existing research identified both technical and nontechnical challenges that hinder adoption of ML in the domain of HRM, we organized our research tasks broadly around the life cycle of data technology adoption. This life cycle involves multiple functional organizations, so we present our findings as a series of tailored reports on different topics. The other reports in this series are:

- *Leveraging Machine Learning to Improve Human Resource Management: Volume 1, Key Findings and Recommendations for Policymakers*, by David Schulker, Matthew Walsh, Avery Calkins, Monique Graham, Cheryl K. Montemayor, Albert A. Robbert, Sean Robson, Claude Messan Setodji, Joshua Snoke, Joshua Williams, and Li Ang Zhang, RR-A1745-1, 2024
- *The Personnel Records Scoring System: Volume 3, A Methodology for Designing Tools to Support Air Force Human Resources Decisionmaking*, by David Schulker, Joshua Williams, Cheryl K. Montemayor, Li Ang Zhang, and Matthew Walsh, RR-A1745-3, 2024
- *Safe Use of Machine Learning for Air Force Human Resource Management: Volume 4, Evaluation Framework and Use Cases*, by Joshua Snoke, Matthew Walsh, Joshua Williams, and David Schulker, RR-A1745-4, 2024
- *Machine Learning–Enabled Recommendations for the Air Force Officer Assignment System: Volume 5*, Avery Calkins, Monique Graham, Claude Messan Setodji, David Schulker, and Matthew Walsh, RR-A1745-5, 2024.

These closely related volumes share some material, including some definitions and descriptions.

The research reported here was commissioned by the Director of Plans and Integration, Deputy Chief of Staff for Manpower and Personnel, Headquarters U.S. Air Force (AF/A1X) and conducted within the Workforce, Development, and Health Program of RAND Project AIR FORCE as part of a fiscal year 2022 project, "Machine Learning Decision-Support Tools for Talent Management Processes." This is one of five related reports originating from the project. The companion reports describe: (1) an overview for the DAF's senior leaders of strategic considerations as it pursues applications of ML to HRM, (2) a case study approach for evaluating the safety of ML systems for HRM, (3) a technical volume on an ML system for scoring officer records, and (4) a conceptual implementation of an ML system for informing officer assignments.

RAND Project AIR FORCE

RAND Project AIR FORCE (PAF), a division of the RAND Corporation, is the Department of the Air Force's (DAF's) federally funded research and development center for studies and analyses, supporting both the United States Air Force and the United States Space Force. PAF provides the DAF with independent analyses of policy alternatives affecting the development, employment, combat readiness, and support of current and future air, space, and cyber forces. Research is conducted in four programs: Strategy and Doctrine; Force Modernization and Employment; Resource Management; and Workforce, Development, and Health. The research reported here was prepared under contract FA7014-22-D-0001.

Additional information about PAF is available on our website: www.rand.org/paf/

This report documents work originally shared with the DAF on September 13, 2022. The draft report, dated September 2022, was reviewed by formal peer reviewers and DAF subject-matter experts.

Acknowledgments

We thank Gregory Parsons, Director of Plans and Integration, Deputy Chief of Staff for Manpower and Personnel, Headquarters U.S. Air Force (AF/A1X), for his support throughout the project. We thank Col Laura King (AF/A1H) and Doug Boerman (AF/A1X). This research benefited greatly from their input and support. We are deeply appreciative of the assistance we received from many personnel, including Lt Col Monique Graham (2022 RAND Air Force Fellows Program). Finally, we thank the many RAND colleagues who helped with this work: principally, but not exclusively, Melissa Baumann, Benjamin Gibson, Lisa Harrington, Ignacio Lara, Nelson Lim, and Miriam Matthews.

Summary

Issue

The Department of the Air Force (DAF) is working to develop and field artificial intelligence (AI) and machine learning (ML) systems for myriad mission areas and support functions, including human resource management (HRM).

Recent developments have improved the access that organizations have to data and analytic tools, opening a wide range of possible ML projects that they could pursue.

Given resource limitations, decisionmakers must choose wisely which projects to pursue among many promising options.

The DAF needs a framework to evaluate the business value, feasibility, and complexity of proposed projects.

Approach

To understand how the DAF can form a balanced portfolio of AI projects for HRM, we reviewed how private-sector organizations evaluate and select such projects. From the review, we arrived at the five-step framework shown in Figure S.1. Broadly, the framework involves evaluating the business value, technical feasibility, and implementation complexity of possible AI and ML projects and forming a portfolio from these evaluations. Each of these steps draws on multiple predefined criteria, which may be assessed using qualitative or quantitative methods. We demonstrate steps of the framework using 19 use cases for applying AI and ML throughout the DAF HRM life cycle.

Notably, this approach does not purport to find the best approach to a business problem. It finds a potentially useful AI approach to addressing a business problem but does not provide a full analysis of alternatives, including non-AI approaches, to address the problem.

Key Findings

- ML techniques such as supervised and unsupervised learning, optimization, language processing, and reinforcement learning are applicable to many HRM functions.
- ML systems can satisfy four HRM objectives: process improvement, performance improvement, enhancing service member opportunities, and enhancing service member motivation. The use cases considered overwhelmingly involved process improvement.
- To be technically feasible, ML systems must have measurable outcomes, relevant inputs, sufficient data, and a suitable algorithmic approach. In the use cases considered here, data

sufficiency was the most common bounding constraint. Additionally, use cases that involve process improvement are more technically feasible.

- The DAF must balance multiple objectives as it builds a portfolio of ML projects for HRM. This includes the relative weight of effort supporting core business functions versus more transformational initiatives. Another consideration is the types of implementation complexity that different initiatives entail.

Recommendations

- **To maximize return on investment, the DAF must use a systematic process to evaluate AI projects for HRM and to build a balanced portfolio.** This entails formulating a business problem, linking it to sources of business value, assessing the feasibility of the AI solution, and evaluating the complexity of implementing it.
- **The DAF should shape an innovation portfolio that includes low-risk/low-reward projects along with higher-risk but potentially transformative ones.** To solidify near-term gains while building a pipeline of potentially more-transformative initiatives, the DAF should allocate about 70 percent of resources to projects that address core HRM processes and 30 percent to those that are more transformative.
- **The DAF should develop a common ML ecosystem to enable rapid creation, comparison, and reuse of ML pipelines, models, and U.S. Department of Defense datasets.** The underlying structure for many of the prediction and decision problems we considered was similar. Thus, methods and models that work for one problem are likely to work for others. A common ecosystem would standardize workflows and enable reuse of ML capabilities across HRM problems.
- **To enable applications of AI to HRM, the DAF must continue to invest in data infrastructure and outcome definitions.** For most use cases we considered, a suitable AI methodology exists. However, for many use cases, suitable inputs, measurable outputs, or both do not currently exist. This limited the feasibility of applying AI for well over half of the HRM problems we considered. To enable applications of AI to HRM, the DAF must continue to invest in data infrastructure and outcome definitions.
- **As the DAF evaluates projects, it must consider the types of technical and nontechnical complexity they entail**. Different types of complexity require different types of human capital and different strategies to overcome.

Figure S.1. Framework for Selecting a Portfolio of AI Projects for HRM

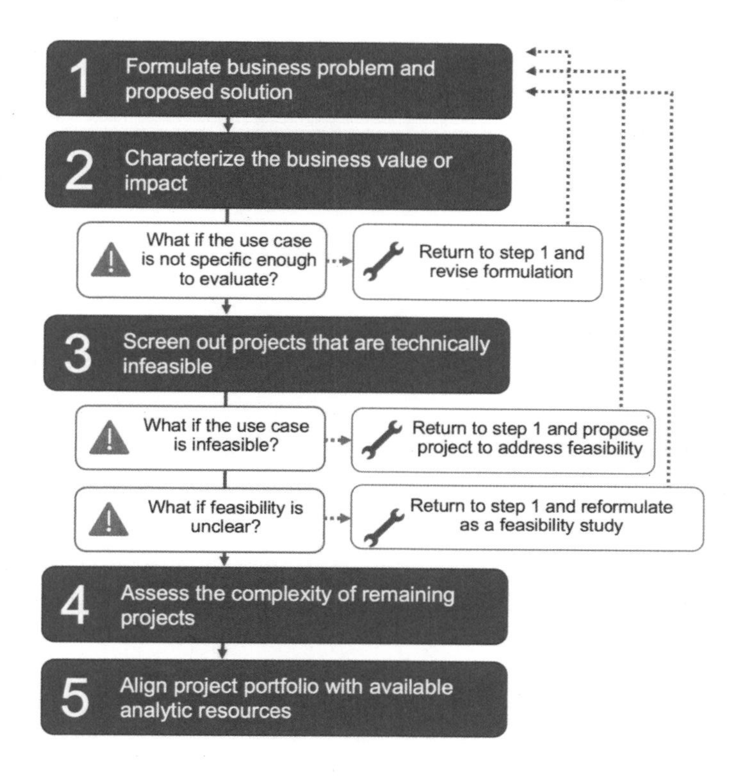

Contents

Figures

Tables

Chapter 1. Introduction

The Department of the Air Force (DAF) is working to develop and field artificial intelligence (AI) and machine learning (ML) systems for myriad mission areas and support functions, including human resource management (HRM). At the same time, the DAF has moved rapidly to create the infrastructure necessary to achieve U.S. Department of Defense (DoD) strategic data goals. These efforts in parallel open the door to a wide range of possible AI projects that could be pursued, but these projects can be difficult to prioritize.

To help HRM decisionmakers get the maximum value out of newly available data capabilities, information technologies, and breakthroughs in the areas of AI and ML, we developed a tailored framework drawn from the research and business literature about prioritizing AI projects. Table 1.1 displays how this framework fits into our series of reports for the DAF regarding uses of ML for HRM.

Table 1.1. Outline of Report Series

Volume Number	Report Title	Report Purpose
1	*Leveraging Machine Learning to Improve Human Resource Management: Volume 1, Key Findings and Recommendations for Policymakers* (Schulker, Walsh, et al., 2024)	Overview for senior leaders
2	*Machine Learning in Air Force Human Resource Management: Volume 2, A Framework for Vetting Use Cases with Example Applications* (Walsh et al., 2024)	Framework for how to prioritize ML projects
3	*The Personnel Records Scoring System: Volume 3, A Methodology for Designing Tools to Support Air Force Human Resources Decisionmaking* (Schulker, Williams, et al., 2024)	Technical report on scoring officer records
4	*Safe Use of Machine Learning for Air Force Human Resource Management: Volume 4, Evaluation Framework and Use Cases* (Snoke et al., 2024)	Case study approach to ensure safety of ML systems
5	*Machine Learning–Enabled Recommendations for the Air Force Officer Assignment System: Volume 5* (Calkins et al., 2024)	ML system to inform officer assignments

NOTE: Current report is highlighted.

Background

The *DoD Data Strategy* summarizes its strategic data goals with the acronym *VAULTIS*, which means that DoD seeks to make its data visible, accessible, understandable, linked, trustworthy, interoperable, and secure.[1] Even before the 2020 publication of the *DoD Data*

[1] DoD, *DoD Data Strategy*, September 30, 2020.

Strategy, the DAF had created a cloud-based VAULT platform (an acronym composed of the first five strategic data goals) to support the "full lifecycle of data exploitation activities."[2] The VAULT has since become part of a "fabric" of federated big-data platforms that, according to DAF officials, has grown increasingly mature over time.[3]

These developments have improved the access that organizations and their analysts have to data and analytic tools, expanding the array of possible projects, including advanced decision support capabilities that use AI and/or ML (AI/ML). In a 2020 interview, the DAF Chief Data Officer aptly summarized the situation that many organizations now face:

> There are a million potential uses for AI/ML, and we've only begun to scratch the surface in what it can do to protect our warfighters, help us make better strategic decisions, help us identify threats early, help us manage or prevent pandemics, and help us fight climate issues. The impact of AI/ML on everyday life has the potential to be huge and positive, but it's our goal, and our responsibility, to make sure that those opportunities can move forward at the speed of mission.[4]

In this context, decisionmakers must choose wisely which projects to pursue among many options that might sound very promising. Identifying and evaluating ideas for new products presents a significant challenge to any organization, as decisionmakers must consider whether the product supports the core business function in a way that other potential solutions cannot, the cost and complexity of developing the product relative to its business value, and any associated development risks.[5] Often, decisionmakers must decide whether to support or reject projects before they have complete clarity about these factors, which further increases the risk that they might unintentionally delay high-value initiatives and fund those that turn out to be "duds."

In this report, we will first describe our tailored framework for building a portfolio of AI projects and then illustrate its utility by applying the framework to a set of potential use cases that span the HRM life cycle. While many of the techniques we illustrate could be used in any HRM setting, we focus on the management of active-duty military personnel in the U.S. Air Force (USAF).

---

[2] Secretary of the Air Force Public Affairs, "Chief Data Office Announces Capabilities for the VAULT Data Platform," October 11, 2019.

[3] Summer Myatt, "Air Force's Data Fabric in Maturation Stage, Officials Say," *GovConWire*, March 30, 2022.

[4] Kathleen Walch, "How the Department of the Air Force Is Driving Forward with AI," *Forbes*, November 14, 2020.

[5] E. Gutiérrez, I. Kihlander, and J. Eriksson, "What's a Good Idea? Understanding Evaluation and Selection of New Product Ideas," in *DS 58-3: Proceedings of ICED 09, the 17th International Conference on Engineering Design*, Vol. 3, *Design Organization and Management*, Design Society, 2009.

What Is HRM?

HRM is the "process of managing an organization's employees," with a view toward meeting the organization's goal.[6] In the USAF, HRM includes a range of functions under the purview of the Deputy Chief of Staff for Manpower, Personnel, and Services. Although not exhaustively, we target primary HRM functions that are broadly applicable to all active-duty airmen. We exclude functions that do not directly relate to management of individual members (e.g., setting end strength or career field strength).

A Life Cycle View of HRM

HRM functions are often depicted as phases in the employee life cycle. An employee enters an organization, grows within it, and eventually leaves, typically to be replaced by another entrant who restarts the cycle. Different HRM functions come into play at different points in this cycle. For the purposes of this report, our view of the life cycle is shown in Figure 1.1. Additionally, we will address "cross-cutting" functions that affect employees in all life cycle phases (though they may have cyclical elements as well).

Figure 1.1. HRM Functions

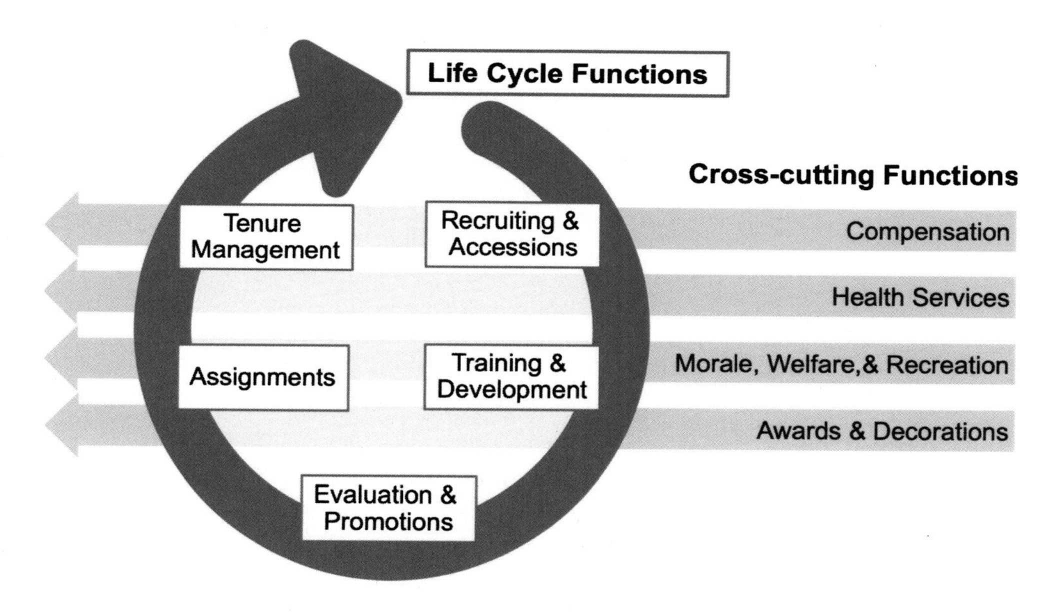

What Are AI and ML?

AI refers to machine-driven task performance typically associated with human intelligence. The current wave of AI applications employs statistical models that use ML methods. *ML* refers

---

[6] Society for Human Resource Management, "HR Glossary," webpage, undated.

to techniques that allow an algorithm to automatically derive functional rules from data, rather than relying on a programmer to specify rules for how the system should function. AI has been the subject of renewed interest in recent years caused by the development of ML tools that might solve some of its oldest problems. These tools include *natural language processing* (NLP), which helps machines to understand human language; *statistical classifiers* and *neural networks*, which help transform patterns found in data into accurate predictions and recommendations; and *network methods*, which help to build connections between concepts or symbols. Though the field of AI has not achieved a machine that can adapt to a wide variety of situations like a human can, it can be very useful when applied to a narrow task set, such as playing chess or *Jeopardy!*, answering specific questions, or making customized recommendations to individuals aligned with their past behavior.

This volume will use the most common methodological vocabulary when describing AI systems, but, for clarity, it is necessary to define some terms up front. Table 1.2 lists these terms with their definitions and a brief example of how the technique might be applied in the USAF context.

Table 1.2. Key Approaches in AI Research

Term	Definition	USAF Example
Supervised ML	Techniques that seek to transform data inputs into a predicted output, based on example pairs of inputs and outputs	Using data on past trainees to create a model that predicts training attrition risk based on trainee characteristics, such as fitness and aptitude
Unsupervised ML	Techniques that seek to transform and organize data inputs without regard to producing a particular output, usually with the goal of understanding or simplifying data inputs	Using an algorithm to identify groups of similar positions for tracking and workforce planning
Reinforcement learning (RL)	Techniques that seek to learn predictive rules using a reward function representing feedback from the system's environment (rather than based on example pairs of inputs and outputs)	Algorithms capable of piloting an F-16 in simulated combat for the Defense Advanced Research Projects Agency's AlphaDogfight Trials were developed through RL[a]
NLP	Cross-cutting term for ML techniques that generate application functionality from models of human language	Using sentiment analysis to automatically classify open-ended survey responses as positive or negative
Optimization	Finding a set of inputs that produces the best possible result from a model	Finding a set of selective reenlistment bonus (SRB) multipliers that achieves retention objectives at the lowest possible cost

[a] Adrian P. Pope, Jaime S. Ide, Daria Mićović, Henry Diaz, David Rosenbluth, Lee Ritholtz, Jason C. Twedt, Thayne T. Walker, Kevin Alcedo, and Daniel Javorsek, "Hierarchical Reinforcement Learning for Air-to-Air Combat, in *2021 International Conference on Unmanned Aircraft Systems (ICUAS)*, Institute of Electrical and Electronics Engineers, 2021.

There is no one-size-fits-all approach to building AI systems. AI necessarily replaces or augments human intelligence in the completion of tasks. In some applications, it replaces a human function completely, such as applying AI to customer service. Google's and Amazon's virtual personal assistants use ML and NLP to enhance AI capabilities and are designed to run without human interference. In other applications, AI provides additional information to help human decisionmaking—for example, recommendation systems for media and automated planners for inventory management and distribution. The applications of AI vary substantially in scope and function, depending on the goals of its developer and whether AI is intended to replace or augment human intelligence.

Each of the approaches described in Table 1.2 requires a different type of data:

- Supervised learning requires training data in which input features are known and outcomes are labeled—for example, whether individuals with different assignment histories were selected for promotion.
- Unsupervised learning requires training data for which input features are known—for example, assignment histories for different individuals.
- Optimization requires a mathematical specification of the environment in which a process takes place and a value function to evaluate the goodness of different potential solutions—for example, a model of which occupations different enlisted personnel can enter, and a way to calculate the cumulative wait time given basic training arrival dates and class start dates. To be useful, the model must provide a good approximation of the environment, yet it must be sufficiently tractable to be solved using optimization techniques.
- RL typically requires a computational specification of the environment in which a process takes place and a reward function to evaluate the immediate and long-term consequences of taking different actions—for example, a workforce simulation to set SRBs to meet validated requirements in the face of economic impacts. To be useful, the simulation must provide a good approximation of the environment, and it must be possible to train an RL agent in a tractable time.
- NLP methods may be applied to written and spoken language data sources.

The implication is that the suitability of a method for a particular problem depends on whether the proper types of data are available. Alternatively, as new types of data become available, additional methods may become applicable.

When considering potential AI approaches for a given problem, an organization must consider whether the objectives of the approach are aligned with the needs of the problem. Additionally, the organization must consider whether the proper forms of data needed for the approach are available.

How Can AI and ML Improve USAF HRM?

After a comprehensive review of HRM in the DAF, the National Academies of Sciences, Engineering, and Medicine recommended that the DAF move toward a more connected system that manages people more "deliberately" and "systematically" through data-driven decisions.[7] The National Academies acknowledged that the DAF already uses data extensively to inform decisions and policies, but current practices create situations in which data use is siloed and uneven, as well as cases in which new approaches could yield new opportunities. The motivation for the National Academies recommendation was that the DAF competes for talent with industry firms, many of which are redesigning their HRM systems to use "data-driven decision making powered by AI" to deliver more-agile experiences that are personalized for their employees.[8]

Industry research shows that many firms invest in goals and strategies for adopting such systems but still fail to realize them. A common theme among these organizations is that they pursue use cases or pilot studies (sometimes pejoratively labeled "pet projects") without centering them on long-term business value.[9] In contrast, firms that successfully adopt the technologies are much more likely to have identified linkages between technologies and business value before undertaking pilot studies.[10] To be successful in the HRM domain, the DAF must also become more adept at developing and vetting potential use cases and prioritizing the ones that produce the most value for the HRM system as a whole.

Framework for Evaluating AI Use Cases in HRM

HRM decisionmakers at different levels of the DAF frequently make decisions or provide inputs into decision processes that allocate analytic resources across potential projects. Conventional wisdom prescribes a formal decision process that begins with strategic goals and project criteria, assembles a list of viable candidates, and prioritizes them using agreed-upon criteria before allocating available resources.[11] In projects seeking to develop new AI software, the description of the project that is relevant to the business case analysis (known as a *user story* in software development) does not provide enough specificity to evaluate the technical level of effort or risk in the project. The latter requires a technical specification of the solution design,

---

[7] National Academies of Sciences, Engineering, and Medicine, *Strengthening U.S. Air Force Human Capital Management: A Flight Plan for 2020–2030*, National Academies Press, 2020.

[8] Amy Wright, Diane Gherson, Josh Bersin, and Janet Mertens, "Accelerating the Journey to HR 3.0: Ten Ways to Transform in a Time of Upheaval," IBM Institute for Business Value, IBM Corporation, 2020.

[9] Tim Fountaine, Brian McCarthy, and Tamim Saleh, "Building the AI-Powered Organization," *Harvard Business Review*, July–August 2019.

[10] Sam Ransbotham, David Kiron, Philipp Gerbert, and Martin Reeves, "Reshaping Business with Artificial Intelligence: Closing the Gap Between Ambition and Action," *MIT Sloan Management Review*, Massachusetts Institute of Technology, 2017.

[11] For a good example of the normative literature on project selection, see Chapter 2 of Randall L. Englund and Robert J. Graham, *Creating an Environment for Successful Projects*, 3rd ed., Berrett-Koehler Publishers, 2019.

which some firms refer to as a *delivery story*."[12] The solution specification is important to making these decisions because it drives cost and risk considerations.

The framework that we describe combines user and delivery stories in a logical flow that helps decisionmakers develop a portfolio of AI projects with high value while accounting for resource constraints. Figure 1.2 presents a high-level visual overview of the framework. Each step in the framework draws on predefined criteria, which can be assessed using qualitative or quantitative methods. The following chapters describe each step in more detail.

Figure 1.2. Framework for Selecting a Portfolio of AI Projects for HRM

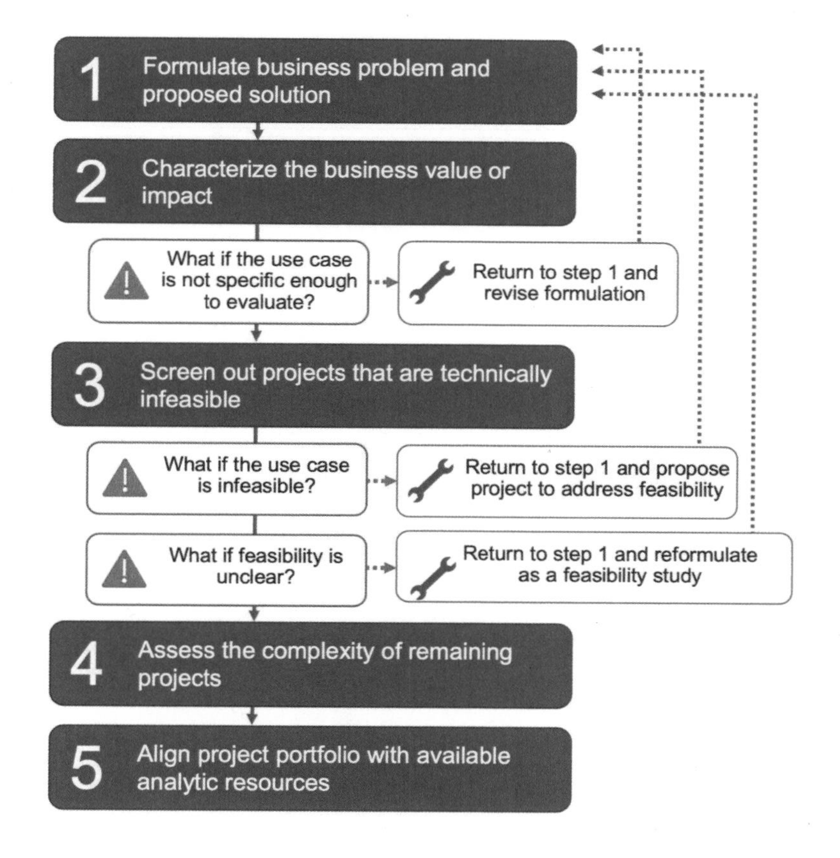

DAF analytic decisionmakers already follow processes for evaluating and prioritizing potential AI projects when they make resourcing decisions. For example, the DAF uses "steering committees" for each research organization that provides analytic support to HRM, and these processes involve panels of stakeholders who rigorously weigh the alignment of each potential effort to strategic objectives. While elements of this framework could be incorporated into such processes, we designed the framework to provide senior leaders with a consistent picture of a

---

[12] Maya Daneva, Egbert van der Veen, Chintan Amrit, Smita Ghaisas, Klaas Sikkel, Ramesh Kumar, Nirav Ajmeri, Uday Ramteerthkar, and Roel Wieringa, "Agile Requirements Prioritization in Large-Scale Outsourced System Projects: An Empirical Study," *Journal of Systems and Software*, Vol. 86, No. 5, May 2013.

wide variety of potential HRM AI applications. When considering alternatives to address a business problem, this approach will increase the likelihood of identifying one or more AI approaches among the competing alternatives and accurately assessing their costs and benefits.

Boundaries of Framework

AI has successfully been applied to HRM in the private sector. Given these private-sector examples, the value of AI in HRM is well established.[13] Additionally, given the numerous vision statements and imperatives issued by the DAF, it clearly intends to use AI to enhance mission areas and support functions—including HRM. This report deals primarily with choosing among data science solutions enabled by AI or ML.

Notably, this approach does not purport to find the very best possible approach to a business problem. It finds a potentially useful AI approach to addressing a business problem but does not provide a full analysis of alternatives, including non-AI approaches, to address the problem. Many HRM challenges may be best addressed in nontechnical ways (e.g., changing manpower requirements) or with technical solutions that do not involve AI or ML (e.g., linking databases to improve enterprise-wide visibility of the workforce or collecting additional data elements). Although this report gives some guidance on when AI solutions are less suitable, the framework we provide is not intended to evaluate the full spectrum of doctrine, organization, training, materiel, leadership and education, personnel, facilities, and policy (DOTMLPF-P) options to meet human capability needs.

In this sense, our approach does not qualify as a *strategic resource management* process.[14] Nonetheless, as the DAF undertakes AI projects, it should be able to articulate capability gaps—such as those expressed in the Air Force Strategic Master Plan and in more-recent DAF and organizational capability gaps lists—that those projects seek to address. In this way, the DAF can seamlessly include the new AI capability in strategic planning, given that it meets the technical requirements specified, and readily adopt the capability if better alternatives are not available. The decision to invest in an AI project is distinct from the decision to deploy an AI system. By filtering out potential projects that do not provide clear business value and that are not aligned with capability gaps, the DAF can increase the chances that the AI projects it invests in lead to AI systems that are ultimately deployed.

---

[13] For reviews, see Pawan Budhwar, Ashish Malik, M. T. Thedushika De Silva, and Praveena Thevisuthan, "Artificial Intelligence–Challenges and Opportunities for International HRM: A Review and Research Agenda," *International Journal of Human Resource Management*, Vol. 33, No. 6, 2022; and Fountaine, McCarthy, and Saleh, 2019.

[14] Robert M. Emmerichs, Cheryl Y. Marcum, and Albert A. Robbert, *An Operational Process for Workforce Planning*, RAND Corporation, MR-1684/1-OSD, 2004.

Purpose of the Report

The purpose of this report, besides describing elements of the selection framework, is to demonstrate its application using a wide range of potential AI projects. Throughout, we assign subjective ratings using our domain and technical expertise, but these are not definitive ratings. DAF subject-matter experts in particular areas should also evaluate the use cases to influence decisionmaking. Nonetheless, our subjective ratings reveal how to apply the framework, and they yield insights into the challenges and opportunities for AI in DAF HRM.

Organization of the Report

The goal of this report is to introduce and demonstrate a framework for selecting a portfolio of AI projects for HRM. To do so, we propose 19 potential use cases. We then demonstrate how to apply the framework using the 19 use cases as inputs. The remainder of this volume is organized as follows:

- In Chapter 2, we describe how to formulate business problems and proposed AI solutions.
- In Chapter 3, we define sources of business value tied to HRM objectives and assess business value for 19 use cases.
- In Chapter 4, we define sources of feasibility and assess the use cases with these.
- In Chapter 5, we enumerate sources of technical and nontechnical complexity for a subset of the use cases considered.
- In Chapter 6, we discuss ways to select portfolios of AI projects for HRM using the outcomes of the earlier steps in the framework.
- In Chapter 7, we summarize findings and offer recommendations.

Chapter 2. Step 1. Formulate the Business Problem and Proposed Solution

The first step in most problem-solving frameworks, technical or otherwise, is to clarify the business problem, objective, or goal that the customer is trying to address.[15] The SMART principle (an acronym for specific, measurable, achievable, relevant and time-bound) for objectives is a way to develop effective formulation statements for future steps in a process.[16] In other words, before one can characterize the business value of the project (Step 2 in Figure 1.2), the objective needs to be specific, measurable, relevant to the strategic HRM objectives of the function, and achievable to avoid being screened out as technically infeasible (Step 3 in Figure 1.2).

In addition to the business problem that the system seeks to address, we recommend including some specificity on the proposed solution and the methods used because this information is necessary for subsequent steps. There is a variety of methods available to formulate problems and solutions systematically and rigorously, such as design thinking techniques.[17] Setting standards for problem–solution formulation at the beginning of the process could save time and resources in the later steps and reduce the risk of malinvestments.

Consider the following two illustrative problem formulations, the latter being intentionally SMARTer than the former:

- Project A: The goal of this project is to use NLP to improve recruiter efficiency and save resources.
- Project B: This project will develop an AI system using supervised ML that ingests recruit data and predicts downstream outcomes. The system will operationalize these predictions for recruiters to reduce their workload by improving prioritization decisions, which ultimately raises the quality of incoming recruits.

Project A contains some elements of specificity in that it states a proposed class of AI techniques (NLP) and a set of business goals (improve efficiency and save resources). However, Project A would be very difficult to evaluate in its current form because it does not contain any information on the business processes it seeks to change or the data sources it will draw on. It is

[15] Pete Chapman, Julian Clinton, Randy Kerber, Thomas Khabaza, Thomas Reinartz, Colin Shearer, and Rüdiger Wirth, *CRISP-DM 1.0: Step-by-Step Data Mining Guide*, NCR Systems Engineering Copenhagen, DaimlerChrysler, SPSS Inc., and OHRA Verzekeringen en Bank Groep, 2000.

[16] Christopher Paul, Jessica Yeats, Colin P. Clarke, Miriam Matthews, and Lauren Skrabala, *Assessing and Evaluating Department of Defense Efforts to Inform, Influence, and Persuade: Handbook for Practitioners*, RAND Corporation, RR-809/2-OSD, 2015.

[17] Robert Stackowiak and Tracey Kelly, *Design Thinking in Software and AI Projects: Proving Ideas Through Rapid Prototyping*, Springer Science and Business Media, 2020.

not specific or measurable enough, and its attainability is unclear. Project B contains enough specificity for an engineer to create a concept for a technical design, and it has multiple linkages to measurable subobjectives. With the second project objective, one could realistically investigate how much prioritization affects recruiter workload and how valuable an increase in recruit quality would be to the USAF. These features put Project B in a good position for Step 2, characterizing the business value.[18]

In the HRM context, one could further winnow the set of possible projects down to a few more-common areas in which AI can feasibly be used. These areas include projects that aim to support *decisions*, such as screening, selecting, or evaluating personnel, with the goal of optimizing HRM *outcomes* with respect to accuracy, consistency, transparency, or cost. The reason these areas tend to be suitable for AI is that the data inputs and HRM outcomes typically are measurable, in line with the core purpose of HRM functions, and associated with clear strategic HRM objectives of interest to the USAF. Areas that are less suitable for AI support systems include ones where each decision is essentially unique, inputs are highly contextualized or not available to a machine, outcomes are subjectively good or bad, or nonintelligent solutions (e.g., *automation*) will suffice.

Application

The purpose of HRM in the DAF is to formulate and administer personnel policies, guidance, programs, and initiatives to develop and sustain the workforce needed to meet strategic military objectives. This occurs in the context of broader constraints, such as pressure to administer HRM functions in a cost-effective manner. AI may allow the DAF to use the capabilities of its workforce more fully, creating strategic advantage. Additionally, AI could automate HRM functions currently performed by humans, netting cost savings.

Companies in the private sector have applied AI to a wide range of HRM functions, including human resource planning, recruitment and selection, training and development, compensation and benefits, and performance management.[19] Although some aspects of these processes differ for the DAF, they overlap considerably with private-sector human resources functions. Drawing from private-sector use cases and combining this with information about DAF HRM processes, we developed the 19 use cases, each with a defined business problem and analytic solution, as described in Table 2.1.

---

[18] Other AI techniques could be used to achieve the objectives described in Project B and could form the basis for a competing set of proposals.

[19] Budhwar et al., 2022.

Table 2.1. AI Use Cases for HRM

Use Case	Business Problem	Analytic Solution
1. Set recruiting resource mix[a]	The USAF must generate roughly 30,000 new enlistment contracts each year. To meet this goal, the DAF must decide how to budget for national advertising and local marketing.	*An AI system could select advertising channels and tailor content to individuals to better meet recruiting goals.* It can analyze national digital usage data to determine the optimal regions, times, and platforms for advertising. The system can also analyze user behavior, such as mouse clicks, to determine the most-effective ads. The AI system can apply optimization techniques using this information to allocate ad money in a cost-effective way.
2. Recruiting chatbot	USAF recruiters must interview, persuade, and shepherd interested applicants, or *leads*, through the enlistment process. This creates significant workload that may place recruiters at risk of burnout.	*AI could reduce recruiter workload by serving as an initial point of contact for interested individuals.* A chatbot can use NLP to retrieve information about career fields, incentive programs, and other aspects of the DAF institution and culture. The chatbot may ask questions to build a lead profile. The system may then recommend occupations and incentives based on the lead's similarity to earlier individuals.
3. Occupational classification	The consequences of career field classification are significant in terms of training success, job performance, and such early-career outcomes as first-term completion and reenlistment.	*An AI system could recommend occupations to individuals to maximize these outcomes.* It can be trained with supervised learning to predict training and early-career outcomes for individuals in different occupations. A complementary system can apply optimization techniques, leveraging the predictive models to assign airmen to occupations to meet USAF needs.
4. Accession date–to-course scheduling algorithm	Enlisted recruits without a guaranteed specialty are given one during basic military training (BMT), However, few training seat vacancies coincide with BMT completion, limiting the USAF's ability to assign those individuals to specialties that they are interested in and qualified for.	*A scheduling algorithm could be used earlier in accession planning to project training seat availability, decide how many contracts to award per specialty, and set BMT ship dates.* An AI system can analyze historical data to form a predictive model of training attrition and course completion times. It can combine information from this predictive model with USAF production goals to determine the time-varying number of training seats needed to accommodate the expected number of individuals accessing and reaching various points in the training pipeline.

Use Case	Business Problem	Analytic Solution
5. Adaptive training	In most USAF training, individuals advance through a fixed curriculum and at a fixed pace. This may waste resources on individuals who, given their prior experiences and abilities, can complete the curriculum more quickly. This may also jeopardize performance of individuals who would benefit from supplementary instruction.	*AI can be used to deliver more-personalized training.* In domains where tasks can be broken down into competencies and where performance can be objectively assessed, AI methods can trace individuals' level of mastery and deliver training that targets the weakest competencies.
6. Developmental education (DE) recommendations	DE selections are based on order of merit, development team (DT) vectors for individual officers, and officer preferences. DE is essential for the individual and the future force, yet selection and assignment is extremely manpower-intensive.	*An AI system could rate officers and recommend DT vectors and DE programs.* It can be trained with supervised learning to predict officer board scores. It can use NLP to extract information from officer performance reports (OPRs) and combine this with information contained in personnel databases. Aside from predicting scores, the system can be trained to recommend DE vectors. Because the number of potential DE programs is so large, unsupervised learning can be used to discover clusters of similar programs.
7. Resource allocation to decrease students awaiting training (SAT) status	Factors like delayed security clearances, medical issues, and training washback and attrition lead to costly inefficiencies in training pipelines. These inefficiencies include longer time to complete training because of delayed course start dates and reduced throughput because of course seats left vacant.	*An AI system could forecast training demand and capacity, and it could program student flows and resources to maximize throughput.* There are many scheduling algorithms. Some use simplifying heuristics, greedy algorithms, dynamic programming, genetic algorithms, and a wide variety of other optimization techniques. These have been used, for example, to create class schedules in schools.
8. Map relationship between assignments and career outcomes	The effects of assignments on future career success and the determination of which assignments are career builders versus career killers are poorly understood. Yet assignment decisions are extremely important for producing the target mix of skill, experience, and diversity.	*An AI system could identify the relationships between early career milestones and future career outcomes.* A Bayesian network, a type of statistical model, can identify which factors influence future career outcomes, and it can be used to simulate the effects of giving individuals different development experiences.
9. Give assignment recommendations	Officers eligible for assignment view available positions and submit preferences through the Talent Marketplace. This may produce assignments that better match officers' needs and preferences. However, the number of positions to consider may be large, and officers may not have complete insight into which positions are most suitable for them.	*An AI system could recommend positions that would maximize development and satisfaction.* It can use content-based approaches to recommend positions like ones the officer previously enjoyed, or a collaborative filter could recommend positions that similar officers liked. Given the large number of positions, the system would use unsupervised learning to find clusters of related positions.

Use Case	Business Problem	Analytic Solution
10. Career coaching	Career coaching is a powerful way to improve service member satisfaction and productivity, but it can be provided to only a limited number of officers and in a limited capacity because it is so time-intensive.	*The USAF could expand the scope and frequency of career mentorship by using AI for career coaching.* An AI career coach can use NLP to retrieve information relevant to a service member's requests. The system can also learn about the service member's interests and goals. The system can use this information to recommend development opportunities.
11. Officer promotion recommendations	The purpose of officer promotion is to select the best-qualified individuals for positions of increased responsibility and authority through a fair and competitive process. However, promotion boards may be susceptible to human error, and they are extremely time-consuming to conduct.	*An AI system could provide recommendations to promotion boards.* It might use a range of analytic techniques to emulate historical board decisions, such as NLP to identify key words or phrases in OPRs and supervised learning to predict outcomes using those words and phrases. Although such a system could reduce workload, it risks replicating historical errors made by human decisionmakers.
12. Change weights of Weighted Airman Promotion System (WAPS) factors	The WAPS determines promotions to the ranks of E-5 and E-6. Eligible candidates receive points according to several criteria. Proper weighting of these criteria is essential to retain the right skill mix in the enlisted force and to advance future senior enlisted leaders.	*An AI system can adjust the weights that the WAPS assigns to each criterion so as to meet the institutional goals and needs of the USAF.* Using optimization techniques to adjust weights can satisfy multiple objectives (e.g., skill mix and diversity) when calculating WAPS scores and determining promotions to E-5 and E-6.
13. Predict future deployability	Accurately forecasting nondeployability rates is critical to contingency planning. By understanding factors that contribute to nondeployability, the USAF can apply programs and policies to increase deployability.	*An AI system could predict future nondeployability.* It can be trained using supervised learning to predict whether an individual will be nondeployable for a variety of reasons. Some of these are within the USAF's control, such as having current medical checks. If an airman's deployment date is known, a rule-based system can confirm that the individual will complete all medical checks before that date.
14. Predict future separation	Accurately forecasting separations is critical to workforce planning. In addition, understanding who is likely to separate and why is essential to delivering counseling and programs to manage retention.	*An AI system could generate more-precise estimates of separation probability using service member and environmental characteristics.* It can be trained using supervised learning to predict whether an airman will separate within a predefined interval of time. These predictions could be used as more-refined retention forecasts or to identify contributors to turnover.

Use Case	Business Problem	Analytic Solution
15. Set high year tenure (HYT)	HYT sets the maximum number of years that an individual can serve in each grade. HYT is an important tool for shaping the enlisted workforce. The USAF can increase or decrease HYT to manage retention in specific workforce segments, as it has frequently done.	*An AI system can forecast future supply and separation rates and program HYT to meet manpower goals.* It can combine predictive models of both separation and the USAF's ability to meet production goals. Given that changes to HYT may have ripple effects on retention at earlier and later years of service, the AI system may use RL to learn how to adjust HYT to achieve immediate and midterm manpower goals.
16. Set SRB levels	The SRB is a cash incentive paid to enlisted members to encourage reenlistments and retention in critical military skills. SRB is an important tool for shaping the workforce.	*An AI system can forecast future supply and separation rates and program SRBs to meet manpower goals.* It can estimate retention elasticities in various enlisted specialties as a function of SRB levels and use those to set SRB levels to meet retention goals. Another approach is to use RL to train an agent, placed in an HRM simulation environment, to adjust SRBs to maintain an optimal state in the face of fluctuations.
17. Recommend morale, welfare, and recreation (MWR) activities	MWR services build resilience in airmen and their families. However, the USAF currently lacks an evidence-informed framework for choosing which MWR services to offer.	*An AI system could track which MWR services are used, link usage to readiness outcomes, and program MWR services to increase resilience.* It can analyze MWR usage to forecast demand by installation and season. Additionally, by linking usage data to service members, the system can analyze relationships between MWR programs or services and outcomes such as retention, health, and readiness. The system can use optimization techniques to determine which MWR programs and services to offer at each location to maximize resilience benefits.
18. Review award recommendations	The USAF issues nearly 100 different awards and decorations, the most of any U.S. military service. These recognize individual and unit accomplishments. Awards and decorations may factor into promotion decisions and service member morale, yet generating and reviewing nominations for awards and decorations is time-consuming.	*An AI system could determine conditions for certain awards and recommend service members for them.* It can use preprogrammed rules to recommend awards and decorations according to recorded criteria (e.g., assignment duration).

Use Case	Business Problem	Analytic Solution
19. Compensation planning	Compensation is critical to recruiting and retention. Yet, information on military compensation and private-sector alternatives is not currently available to policymakers, supervisors, or airmen.	*An AI system could combine data on total compensation, replacement costs, and external data to benchmark compensation levels. The system could be used to counsel airmen, help policymakers to set bonus policies, and help Congress decide how to structure pay.* It can combine sources of information to set compensation at competitive levels by career field to meet recruiting and retention goals. Given the dynamic nature of HRM and of the economy, the AI system may use RL to train an agent, placed in an HRM simulation environment, to adjust compensation to maintain an optimal state in the face of fluctuations.

[a] David Knapp, Bruce R. Orvis, Christopher E. Maerzluft, and Tiffany Berglund, *Resources Required to Meet the U.S. Army's Enlisted Recruiting Requirements Under Alternative Recruiting Goals, Conditions, and Eligibility Policies*, RAND Corporation, RR-2364-A, 2018. This table also features information from Budhwar et al., 2022.

Each use case draws on one or more types of AI. Table 2.2 crosswalks use cases to the types of AI central to each.

Table 2.2. Types of AI Central to Each Use Case

Use Case	Supervised Learning	Unsupervised Learning	RL	NLP	Optimization
Set recruiting resource mix	X				X
Recruiting chatbot	X	X		X	
Occupational classification	X				X
Accession date–to-course scheduling algorithm	X				X
Adaptive training	X				X
DE recommendations	X	X		X	X
Resource allocation to decrease SAT status	X				X
Map relationship between assignments and career outcomes	X				
Give assignment recommendations	X	X			
Career coaching	X	X		X	
Officer promotion recommendations	X			X	
Change weights of WAPS factors	X				X
Predict future deployability	X				
Predict future separation	X				
Set HYT	X		X		
Set SRB levels	X		X		

Use Case	Supervised Learning	Unsupervised Learning	RL	NLP	Optimization
Recommend MWR activities	X				X
Review award recommendations	X			X	
Compensation planning	X		X		

The patterns of usage support several findings.

- *Supervised learning* played a role in *all* use cases. In some cases, the contribution of supervised learning is to predict the primary outcome of interest (e.g., training completion or board scores). In other cases, the contribution of supervised learning is to forecast future events, conditions, or needs that the AI system must account for before reaching a decision (e.g., future production and separations).
  - There is a distinction between predicting future outcomes and shaping them. Most types of ML models are suitable for the former, given that the primary goal is to make accurate predictions (e.g., forecasting retention). Causal models are more suitable for the latter, given that the primary goal is to identify what to change to shape future outcomes (e.g., increasing retention).
  - Relatedly, there is a distinction between predicting future outcomes and understanding past ones. The former places a premium on prediction accuracy, while the latter places a premium on explainability.
- *Optimization* is the next most common approach. In some cases, optimization is used to allocate monetary resources to achieve a goal (e.g., meeting recruiting quotas). In other cases, optimization is used to arrive at the globally best solution for the most individuals (e.g., assigning individuals to training seats).
- *NLP* is used in several case studies. Some use natural language generation to enable communication between humans and the AI system (e.g., interacting with leads). Others use natural language understanding to extract information from text sources (e.g., OPRs).
- *Unsupervised learning* is also used in several case studies. The primary role of unsupervised learning is to group large numbers of options (e.g., assignments or DE programs) into a more tractable number of sets.
- *RL* is used in some case studies. RL may be favored for problems that involve learning sequences of decisions, for which some outcomes are experienced in the future (e.g., HYT, SRB, and compensation planning), and for problems with complex environments. Optimization methods may not be computationally feasible in these cases.

Summary

The first step in the framework is to formulate a business problem and a proposed solution in enough detail to permit evaluation. In this chapter, we formulated 19 AI use cases for HRM. Collectively, the use cases demonstrate the potential to integrate AI into most aspects of HRM. In addition, the use cases demonstrate that a variety of AI approaches are available and potentially useful in meeting HRM needs.

Chapter 3. Step 2. Characterize the Business Value or Impact

The second step in the framework is to take a proposed project and characterize the business value or impact. While decisionmakers might prefer purely quantitative metrics of each project's return on investment (ROI), such information is not always available at the proposal stage. The main purpose of this step is to consider the types of business value that each project seeks to generate and make a credible quantitative or qualitative statement of the contribution of the project to one or more HRM objectives. The more credible the business value claim, the more persuasive it is likely to be to decisionmakers in the resourcing process.

The business literature suggests that many organizations see disappointing results from investments in AI tools, partly because they do not have strong ties between their AI projects and clear business value.[20] Speaking specifically about using AI to support HRM, researchers at IBM state the following:

> HR practitioners should have a direct line of sight from AI applications to the outcomes AI will produce and the associated ROI that occurs in the business. Establishing the expected connection between AI and its return should occur before the AI application is implemented.[21]

Though the USAF faces the significant limitation that financial performance metrics do not exist for its "business," there are other ways to link an AI decision-support tool to credible value for USAF HRM. Extending previous frameworks linking HRM to organizational outcomes, Table 3.1 defines four objectives that AI decision-support tools for HRM may seek to address.[22] The table also lists key metrics to evaluate whether the AI system meets those objectives.

---

[20] Fountaine, McCarthy, and Saleh, 2019.

[21] Nigel Guenole and Sheri Feinzig, *The Business Case for AI in HR, with Insights and Tips on Getting Started*, IBM Smarter Workforce Institute, IBM Corporation, 2018.

[22] Kaifeng Jiang, David P. Lepak, Jia Hu, and Judith C. Baer, "How Does Human Resource Management Influence Organizational Outcomes? A Meta-Analytic Investigation of Mediating Mechanisms," *Academy of Management Journal*, Vol. 55, No. 6, 2012.

Table 3.1. HRM Objectives and Key Metrics for Evaluating Business Value

Objective	Definition	Key Metrics
Process improvement	Designed to improve the efficiency and effectiveness of HRM processes and decisions	Organizational: efficiency, workload, costs Individual: Days not in training
Performance improvement	Designed to improve individual (or unit) performance; increases knowledge, skills, and abilities (KSAs)	Organizational: Air Force Specialty Code training success; average performance evaluations Individual: officer and enlisted performance reports; training grades
Motivation enhancement	Designed to increase motivation of airmen; includes developmental performance management, compensation, rewards, benefits, promotions, job security	Organizational: recruitment, retention, end strength Individual: satisfaction, commitment, engagement
Opportunity enhancement	Designed to empower airmen to use skills and motivation to achieve organizational objectives; includes flexible job design, work teams, employee involvement, information sharing	Organizational: commander feedback; readiness metrics Individual: allocating job tasks that fit person's skills and interests

SOURCE: Features information from Jiang et al., 2012.

Figure 1.2 offers alternatives for AI proposals that are screened out because they do not have clear business value. If the AI system does not have clear business value, the project advocate can reformulate the proposal to make clear the objectives that the system will address. Alternatively, if a strong case cannot be made, the proposal may be set aside. Screening for business value ensures that the intended benefit of the system is articulated to decisionmakers *before* they support a project with resources.

Application

Seven subject-matter experts with knowledge of DAF HRM processes evaluated the 19 use cases described in Chapter 2. Raters evaluated whether each use case met the HRM objectives described in Table 3.1 using a three-point scale: (1) *low*—the use case is unlikely to address the HRM objective; (2) *medium*—the use case is not specifically designed to address the objective, yet it may do so indirectly; and (3) *high*—the use case is specifically designed to address the HRM objective, yielding four distinct scores for each use case.

Ratings

At this stage, business value should be evaluated according to the potential of a use case to meet a particular objective or the cumulative utility across objectives. Figure 3.1 shows average rater scores for each HRM objective and use case. Larger values denote stronger perceived alignment. For example, overall, the 19 use cases appear to have strongest alignment with the objective of process improvement, and weakest alignment with the objective of enhancing opportunity.

Figure 3.1. Ratings of HRM Objectives Achieved by Each Use Case

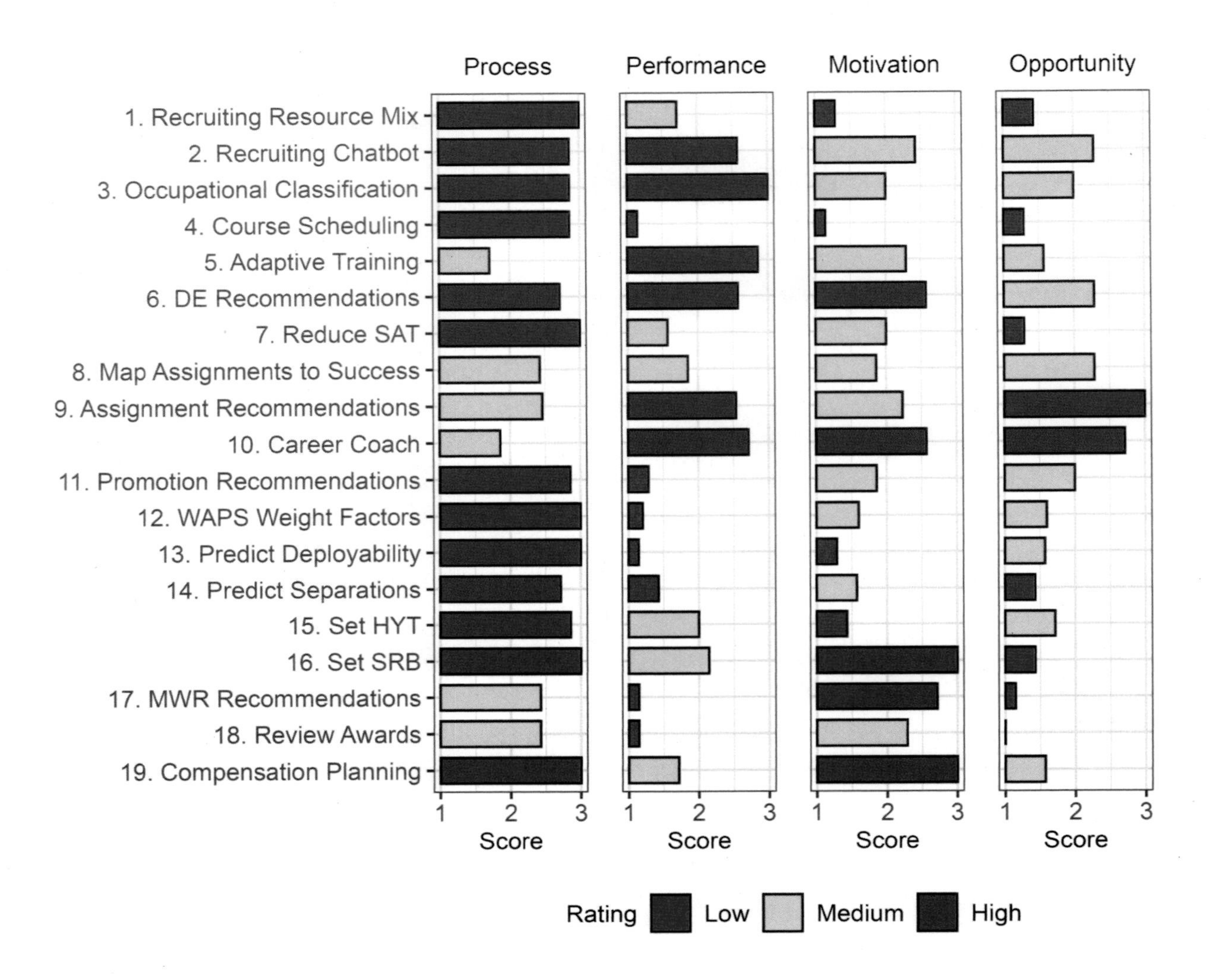

We converted the average ratings shown in Figure 3.1 into category labels corresponding with low, medium, and high alignment. To do so, we treated average scores greater than or equal to 2.5 as reflecting high alignment, average scores greater than or equal to 1.5 and less than 2.5 as reflecting medium alignment, and average scores less than 1.5 as reflecting low alignment. The colors of the bars correspond with these category labels.

At this level of aggregation, alignment varied by HRM objective. Process improvement was scored as high for 68 percent of use cases versus just 32 percent for performance, 26 percent for motivation, and 11 percent for opportunity. The implication is that the AI use cases overwhelmingly favored process improvements.

Alignment also varied by use case. Seventeen use cases were scored as high for one or more HRM objectives. Of those, two were scored as high for three objectives, and five were scored as high for two objectives (Table 3.2). The implication is that some use cases may be especially attractive in terms of business value.

Table 3.2. Use Cases with Highest Business Value

Number of Significant Sources of Business Value	Use Case
Three	6. DE recommendations 10. Career coaching
Two	2. Recruiting chatbot 3. Occupational classification 9. Give assignment recommendations 16. Set SRB levels 19. Compensation planning

Our analysis of the business value rating reveals the following.

- Most use cases directly or indirectly improve processes. Some improvements are tactical in nature. For example, a course scheduler would reduce the number of personnel needed to generate schedules; more-effective scheduling would reduce training time and increase throughput. Importantly, this example uses AI to enhance an existing process. Others entail more-strategic change. For example, it is not currently feasible to deliver career coaching to all officers. An AI-enabled career coach would create new processes for delivering developmental recommendations to all officers regardless of rank.
- Many use cases improve performance. They do so by developing the abilities of individuals (e.g., adaptive training), by assigning individuals to specialties or positions that leverage their unique abilities (e.g., occupational classification), or by retaining individuals with high-value skill sets (e.g., setting SRB levels).
- Some use cases increase motivation. The primary benefit to the USAF is that use cases that increase motivation may increase satisfaction, thereby increasing retention in key workforce segments (e.g., compensation planning). A secondary benefit is that increasing satisfaction may increase performance (e.g., adaptive training).
- Few use cases increase opportunity. These primarily involve allowing individuals to discover high-value opportunities that they otherwise would not (e.g., assignment recommendations or career coach). Some considerations that would modulate the effectiveness of these use cases are whether service members would have flexibility to pursue new opportunities identified by the ML model, and whether the ML model would identify opportunities in an equitable way.

Reformulation

Rater agreement was generally high, but there were significant discrepancies for some combinations of use case and HRM objective. For example, 43 percent of raters said that a career coach had low relevance to process improvement, and 57 percent said it had high relevance. When such disagreement exists, it may signify that the business problem and analytic solution are poorly defined. These use cases may be returned to Step 1 (Figure 1.2) for formulation. Alternatively, disagreements may indicate that the use case has an additional source of business value that is not at once clear to all raters. In these cases, raters should be allowed to adjust their ratings following a consensus meeting discussing the reasoning for their ratings, and the project description should be revised to clarify its intent.

Estimating Order of Magnitude Effects

Estimating order of magnitude effects will be specific to each proposal and, depending on the HRM objectives under consideration, divergent across the set. For example, the cost savings from two AI systems that reduce the number of personnel needed to perform an HRM function can be compared with one another, but not to the proficiency gains produced by a third system that delivers adaptive training.

Modeling and simulation, business case analysis, and implementations of demonstration projects are some of the tools available to quantitively estimate business value. However, at this stage of evaluation, seasoned judgment is required to infer order of magnitude effects.

Summary

A key step to delivering high-impact use cases is to articulate business value in advance. At this stage of analysis, the sources of value are plausible but unproven. A use case may be aligned with one of four HRM objectives: process improvement, enhancing skills and performance, enhancing motivation, and enhancing opportunity. Linking use cases to these objectives ensures that they have potential business value and, in later steps of the decision framework, may permit selection of a portfolio of projects that address the right mix of objectives.

Across the 19 use cases, we found that the potential value of AI overwhelmingly involves process improvement. In addition, we found that certain use cases contributed to three or more of the HRM objectives. Future quantification of the order of magnitude benefits would permit identification of use cases with the highest business value.

Chapter 4. Step 3. Screen Out Projects That Are Technically Infeasible

At this stage of the process, project advocates would have put forward a set of projects and their best possible statement of the business value from the perspective of the functional stakeholders. The third step in the proposed framework is to ensure that each project meets a set of minimum feasibility conditions. These conditions are not a guarantee of success; rather, they exist to test viability and decrease the potential risk of failing to fulfill the intended objectives. Revealing and informing these risks is an important objective of a demonstration project. In Chapter 5, we suggest that a project combining high risk (as a function of complexity) with high business value has an important place in a portfolio of demonstration projects.

Though the USAF cannot eliminate risk from projects at the time of selection, there are ways to link an AI decision-support tool to its feasibility. Table 4.1 lists four criteria that a system must meet to be feasible.

Table 4.1. Criteria for Determining Feasibility of AI Solution

Criterion	Definition	Key Metrics
Measurable outcomes	AI decision-support systems will not be able to improve decisions without a way to differentiate good outcomes from bad outcomes. Thus, the problem must have measurable outcomes.	• Manpower, time, money • Training time and completion • Unit readiness reporting • Accession, recruiting, and other programmatic goals
Relevant inputs	AI systems will not be able to deliver their proposed value if the input factors are not measurable and thus able to be included in an ML model. In other words, relevant inputs must be known and measurable.	• Identifiable constructs linked to measurable elements
Data sufficiency	Without data of sufficient quantity and quality, an ML system will not be able to learn an adequate model. Thus, data sufficiency is a necessary condition.	• Amount of data • Completeness and accuracy of data
Methodological suitability	Although AI systems have advanced rapidly, there remain many tasks and challenges that such systems are not well suited for. Thus, the process should screen for whether an appropriate methodology exists, and whether that is the one included in the problem, as formulated.	• Prior demonstrations to related problems

Figure 1.2 offers alternatives for AI proposals that are screened out because they are not technically feasible. If the AI system is not feasible, the project advocate can reformulate the project around the objective of addressing the feasibility issues that caused the project to be screened out. Furthermore, if the feasibility of the project is indeterminate, then project advocates can reformulate the project as a feasibility study.

Application

Three subject-matter experts with knowledge of DAF HRM processes and AI evaluated the 19 use cases described in Chapter 2. Raters evaluated the extent to which each use case met the feasibility criteria described in Table 4.1 using a three-point scale: (1) *low*—infeasible, (2) *medium*—some feasibility concerns, and (3) *high*—feasible.

Ratings

At this stage, feasibility should be evaluated according to the limiting factor or the cumulative feasibility barriers. Figure 4.1 shows average rater scores for each feasibility criterion and use case. Larger values denote greater technical feasibility. For example, overall, the 19 use cases appear to have highest technical feasibility in terms of the existence of a suitable methodology and lowest feasibility in terms of the existence of sufficient data.

Figure 4.1. Ratings of HRM Objectives Achieved by Each Use Case

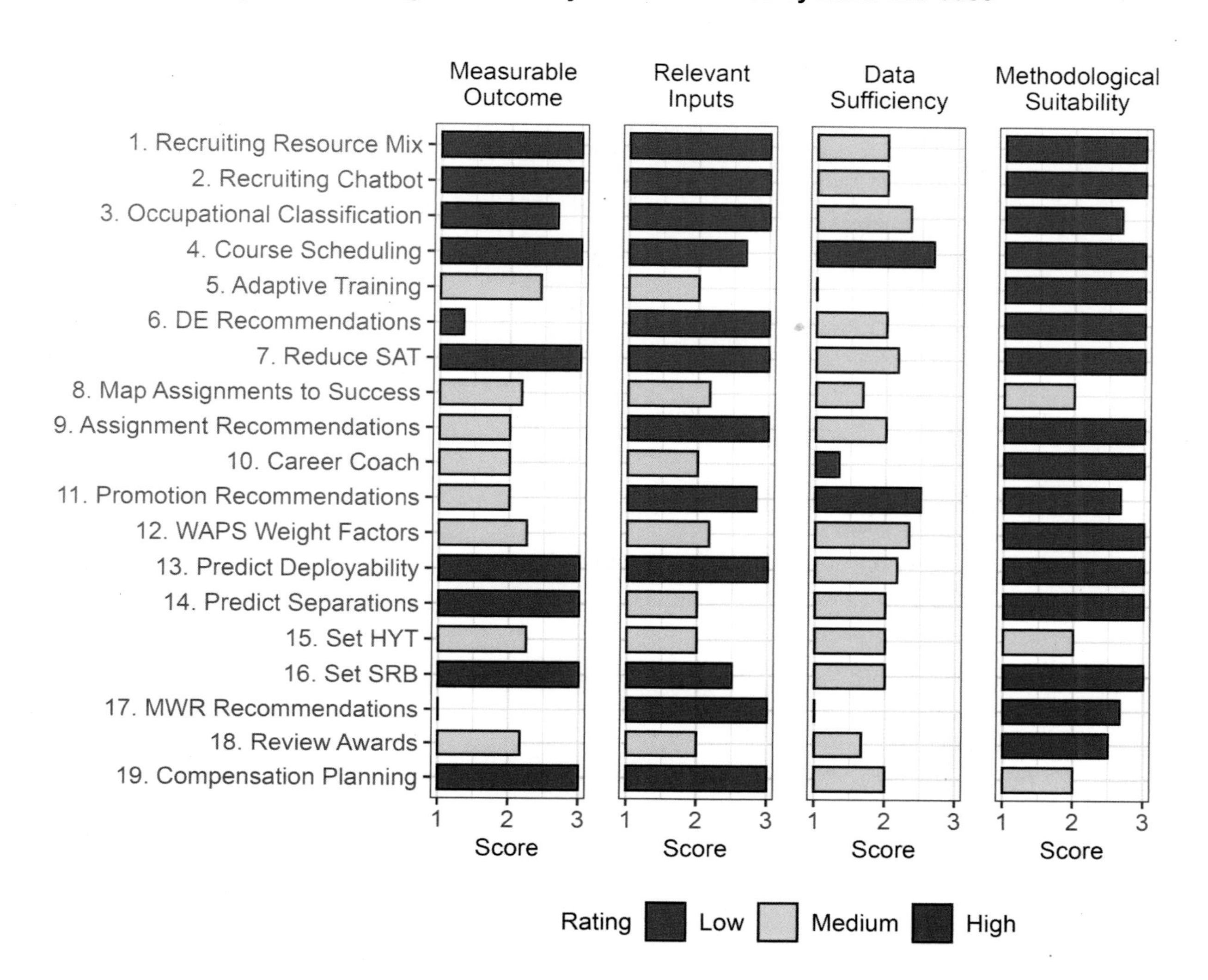

We converted the average ratings shown in Figure 4.1 into category labels corresponding with low, medium, and high technical feasibility. To do so, we treated average scores greater than or equal to 2.5 as reflecting high feasibility, average scores greater than or equal to 1.5 and less than 2.5 as reflecting medium feasibility, and average scores less than 1.5 as reflecting low feasibility. The colors of the bars correspond with these category labels.

At this level of aggregation, feasibility varied by technical criterion. Methodological suitability was scored as high for 84 percent of use cases, versus just 63 percent for relevant inputs, 47 percent for measurable outcomes, and 11 percent for data sufficiency. The implication is that suitable algorithmic approaches exist for most case studies. The primary limiting factor is whether the data needed for the algorithm are available or could even be gathered.

Feasibility also varied by use case. One use case was scored as high for all technical criteria and seven were scored as high for all but one technical criterion (Table 4.2). The remaining 11 use cases were scored as high for two or fewer technical criteria. The implication is that, on account of technical feasibility, certain use cases may be especially likely to fail.

Table 4.2. Use Cases with Highest Technical Feasibility

Number of Feasible Technical Dimensions	Use Case
Four	4. Accession date–to-course scheduling algorithm
Three	1. Set recruiting resource mix 2. Recruiting chatbot 3. Occupational classification 7. Resource allocation to decrease SAT status 11. Officer promotion recommendations 13. Predict future deployability 16. Set SRB levels

Our analysis of technical feasibility ratings revealed the following.

- A suitable methodology exists for most use cases. Established approaches for optimization, supervised learning, and unsupervised learning that have been applied in many areas, including HRM, can also be used by the USAF (e.g., course scheduling, predicting separation, and assignment recommendation). For dynamic problems with long-range dependencies between current and future decisions (e.g., setting HYT), suitable methods are not as well-established.
- Relevant inputs are available for many use cases. The USAF can collect data on cognitive aptitudes, psychological makeup, personality, and work interests of service members. These inputs underlie such use cases as occupational classification and assignment recommendation. The USAF also has data on programmed end strength and career field strength, which are needed to set accession, training, and retention goals. Notwithstanding this coverage, it is relatively more difficult to measure service member attitudes and job satisfaction because of their subjective nature and the potential for them to be affected by social desirability bias.[23] In addition, relevant inputs may not yet be available for decisions that occur very early in an individual's career. Finally, for future-focused decisions (i.e., *strategy*), the USAF lacks precise information about future needs and circumstances.
- Measurable outcomes are defined for some use cases. Ones that involve reducing manpower, time, and money to accomplish an existing process are well suited for measurement (e.g., course scheduling). Additionally, ones that involve reducing training time or meeting accession and retention goals are also well suited for measurement (e.g., recruiting resource mix, reducing SAT, and setting SRB). Other outcomes, such as job performance and resilience, are more difficult to measure.[24] Satisfaction can be measured,

---

[23] Ivar Krumpal, "Determinants of Social Desirability Bias in Sensitive Surveys: A Literature Review," *Quality and Quantity*, Vol. 47, No. 4, 2013.

[24] Sarah O. Meadows, Stephanie B. Holliday, Wing Y. Chan, Stephani L. Wrabel, Margaret Tankard, Dana Schultz, Christopher M. Busque, Felix Knutson, Leslie A. Payne, and Laura L. Miller, *Air Force Morale, Welfare, and*

but the DAF does not have this information available yet and would need to identify appropriate ways to gather data and measure it.

- Sufficient data are available for only some use cases. To satisfy this criterion, relevant inputs and measurable outcomes must both be available. The data must be of reasonable quality, which may not be the case if they are manually entered and prone to error. Moreover, *enough* data must be available to use the algorithmic approach. The USAF is made up of over 600,000 officers, enlisted personnel, and civilians, for whom decades' worth of transactional and administrative data are available. Even so, this is only a fraction of the data available from Amazon or Netflix users, for example. In addition, due to the changing nature of USAF policy, needs, and external conditions, historical data may lose currency. Finally, the total force is made up of hundreds of personnel subcategories. Once the data are fractionated, far less information remains for each subcategory.

Relationship Between Objectives and Feasibility

To determine whether technical feasibility varies based on the HRM objectives that a use case addresses, we computed cross-correlations between the average ratings for HRM objectives and technical feasibility across the 19 use cases (Table 4.3). Use cases that improved a process were rated as having higher technical feasibility in terms of measurable outcomes, relevant inputs, and data sufficiency. Conversely, use cases that enhanced motivation were rated as having lower technical feasibility in terms of data sufficiency and, to a lesser extent, measurable outcomes. The implication is that use cases that primarily involve process improvement tend to have higher technical feasibility.

Table 4.3. Correlations Between Ratings for HRM Objective Alignment and Technical Feasibility

	Technical Feasibility			
HRM Objective	**Measurable Outcomes**	**Relevant Inputs**	**Data Sufficiency**	**Methodological Suitability**
Process improvement	0.46*	0.50*	0.79**	–0.05
Performance improvement	–0.06	0.01	–0.26	0.12
Motivation enhancement	–0.31	0.12	–0.49*	–0.09
Opportunity enhancement	–0.27	0.09	–0.01	0.05

** $p < 0.01$; * $p < 0.05$

Reformulation

Figure 1.2 contains two branches for projects screened out due to technical feasibility. If a use case has low technical feasibility, it may be reformulated as a study to increase technical feasibility. For example, compensation planning involves long-range dependencies between

---

Recreation Programs and Services: Contribution to Airman and Family Resilience and Readiness—Appendixes, RAND Corporation, RR-2670-AF, 2019.

today's policies and the makeup of the workforce for many years into the future. RL is an effective approach for learning to control dynamic systems, but it has not yet been demonstrated for HRM. A study to increase technical feasibility could develop a simulation of the HRM life cycle and show how an RL agent can be trained to learn effective compensation policies.

If a use case has *unknown* technical feasibility, it may be reformulated as a study to assess technical feasibility. For example, MWR programs and services are intended to increase the resilience of service members and of their families. Although there are indirect measures of resilience, such as suicide, depression, and substance abuse, there are not direct measures of this construct for HRM. A study to assess technical feasibility could survey approaches for measuring resilience and determine whether they are suitable for evaluating MWR programs and services.

In both cases, the nominal business value of the new project will likely be less than the original, but screening for feasibility ensures that realistic benefits and costs are brought before decisionmakers before they devote resources to support a project. In addition, overcoming feasibility challenges may create new, future sources of business value.

Summary

To increase the likelihood that AI use cases are suitable, they should be screened to ensure that they meet minimum feasibility conditions. A use case may be evaluated for technical feasibility according to four criteria: measurable outcomes, relevant inputs, data sufficiency, and methodological suitability. Evaluating use cases along these dimensions helps to avoid pursuing projects with low technical feasibility.

Across the 19 use cases, we found that suitable methodologies exist for most, and that data sufficiency is often the limiting factor. In addition, we found that certain use cases are likely to fail due to the significant concerns involving multiple technical criteria.

Chapter 5. Step 4. Assess the Complexity of Remaining Projects

The fourth step in the proposed framework calls for an assessment of each proposed AI system's complexity, or "the scope of the challenge that you face in delivering the project."[25] This goes beyond the previous step in asking whether developing and fielding a particular AI system is practically feasible. A candidate AI system's complexity could affect either the cost of development or the likelihood that the project will fall short of generating the expected business value. Complexity includes technical dimensions, such as the level of development skill or effort required to build the AI system and the underlying technology level that the system calls for. However, it can also include nontechnical factors, such as social considerations that affect the adoption of the AI system and ease of implementation.[26]

There are many different potential complexity dimensions that decisionmakers might consider. In applying the framework to real-world use cases, we will aim broadly and use *technical complexity* to refer to the scale of the functional challenge in developing the system (which will encompass issues with accessing and transforming data, implementing a suitable model, and embedding it in a decision-support system). We will use *nontechnical complexity* to refer to the scope of challenges related to using the tool in practice, including the policy structures governing the decisions and data use, risks of unintended consequences, etc.

Note that the goal of this step is not to deprioritize complex projects, as many of the most complex projects might also have the greatest business value. The goal is to assess the complexity of each candidate so that decisionmakers can take a portfolio approach that weighs value and complexity, as we describe in the following section.

Application

To demonstrate some of the considerations that may emerge from this step, we expanded upon five of the use cases described in Table 2.1 that reflect different levels and types of complexity. In order of the phases of the HRM life cycle, the use cases were occupational classification, adaptive training, assignment recommendations, promotion recommendations, and predicting future separations. The appendix contains complete descriptions of the use cases.

---

[25] Nigel Guenole, Jonathan Ferrar, and Sheri Feinzig, *The Power of People: Learn How Successful Organizations Use Workforce Analytics to Improve Business Performance*, Pearson Education, 2017.

[26] Guenole, Ferrar, and Feinzig, 2017.

Ratings

Table 5.1 summarizes all sources of technical and nontechnical complexity that the project team identified across the five use cases. Overall, the most common source of technical complexity involved the need to access data from multiple government-owned and proprietary administrative systems. A related source of complexity is the need to integrate the AI system with existing IT systems (i.e., the *technology ecosystem*), both to access data and inputs, and to provide outputs to human decisionmakers or other downstream processes. Given the patchwork commercial and government-owned IT systems that underlie DAF HRM, these sources of technical complexity may be significant. A final common source of technical complexity was the difficulty of measuring outcomes.

One common source of nontechnical complexity involves upskilling the workforce to build, deploy, and sustain AI systems. The application of AI to HRM functions also drives the need for additional education and training to allow operators to use the system properly. Another source of nontechnical complexity involves properly handling personnel data. For example, an AI system may use medical or legal data. Given the potentially sensitive nature of these data, the USAF must take additional steps to safeguard individuals' privacy.

Table 5.1. Examples of Technical and Nontechnical Complexity

Type	Description	Occupational Classification	Adaptive Training	Assignment Recommendations	Promotion Recommendations	Predicting Separation
Technical	Data must be accessed from multiple different administrative systems	+		+	+	+
	Data are not available in machine-readable format				+	
	The process uses unstructured or semistructured data			+	+	
	New data sources that may improve predictions are only selectively available	+		+		
	Historical screening criteria have restricted the ranges of input variables	+				
	Policies that affect the process have changed, devaluing historical data	+		+	+	
	The ML system must make decisions for novel cases	+		+		
	The size of the decision space is large			+		
	Outcomes are difficult to measure	+	+	+	+	
	Outcomes are influenced by multiple factors	+		+		
	Outcomes of a decision extend far into the future	+		+	+	
	Outcomes have little variability				+	+
	The system requires changes to the technology ecosystem		+			
Nontechnical	The process involves multiple stakeholders	+		+	+	
	Stakeholders have competing interests	+		+	+	
	The system requires a significant cultural shift		+			
	Policies and adjacent processes must be changed to support the system	+	+			
	Data used by the system raise privacy concerns	+	+	+	+	+
	Operators must be trained to use the system	+		+	+	
	The workforce lacks technical skills to build, evaluate, and deploy the system	+	+	+	+	+
	The system faces competition with an entrenched competitor					+

Although it is tempting to rate overall complexity of a use case according to the number of sources of complexity that it entails, this could be misleading. For example, although adaptive training had the second fewest sources of complexity, ones that it did have, such as transitioning any count-based measures of readiness to proficiency-based, would be difficult to overcome. Thus, expert judgment must be applied to the outcomes of the complexity analysis. The red shapes in Figure 5.1 show where we place the five use cases on the spectrum of technical and nontechnical complexity (and the green shapes, discussed in the next section, show where we place the five use cases after applying steps to limit technical and nontechnical complexity).

Our assessment is that adaptive training, the top-right red square, is the most complex on both dimensions. However, four of five proposed use cases are of medium-to-high complexity in both their technical and nontechnical aspects.

Figure 5.1. Complexity Assessment for Use Cases

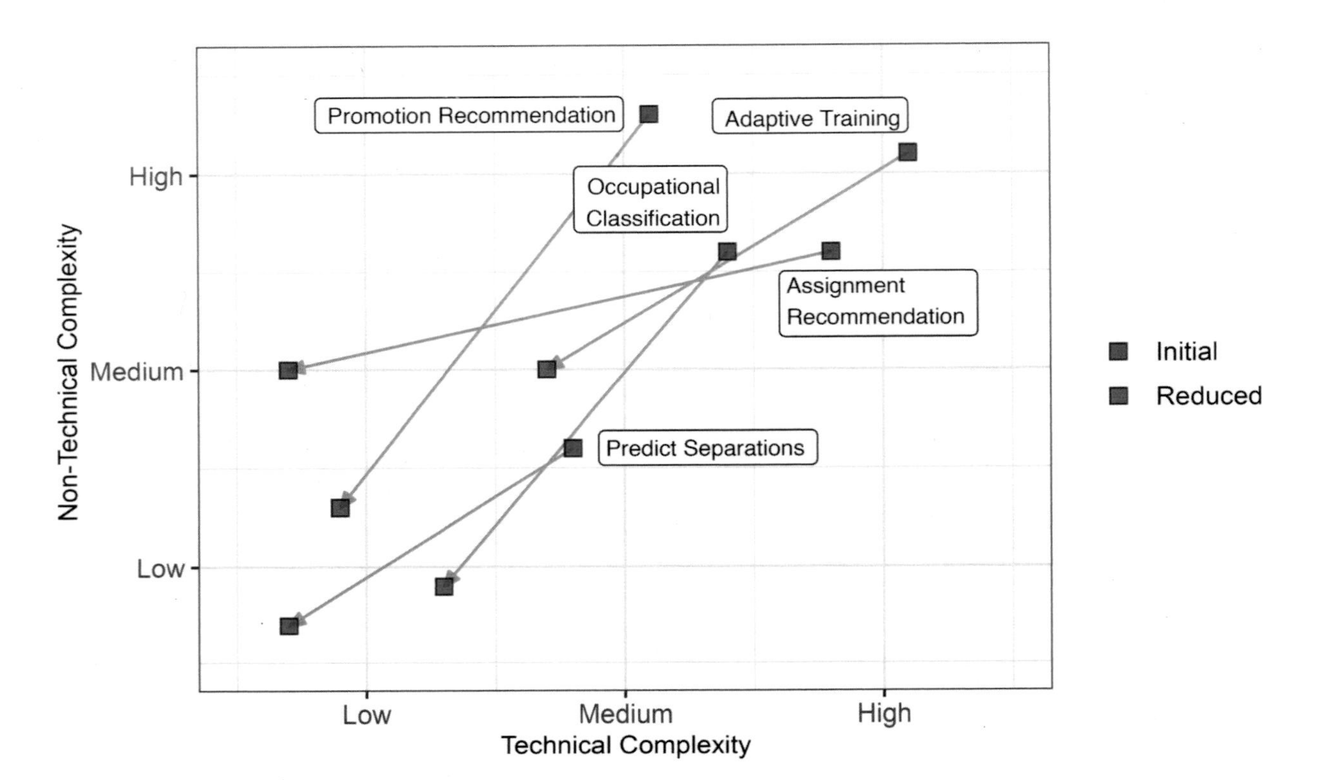

Reducing Project Complexity

Importantly, there are steps the DAF can take to limit the moderate-to-high technical and nontechnical complexity of all use cases. Some of these steps are described in Table 5.2. In general, to reduce *technical* complexity, the DAF can limit the initial scope of applications outcomes, skills, or workforce segments that are most amenable to ML approaches. In addition, the DAF can collect and make key data sources readily available. To reduce *nontechnical* complexity, the DAF can limit the influence of the ML system—for example, by providing recommendations rather than automating decisions. In addition, the DAF can focus initial applications on lower-stakes processes, such as developmental boards rather than promotion boards.

Table 5.2. Actions to Reduce Technical and Nontechnical Complexity

Use Case	Technical Complexity	Nontechnical Complexity
Occupational classification	• Focus on measurable outcomes, such as training completion • Focus on hard-to-fill specialties	• Deliver recommendations rather than automating decisions • Begin by focusing on individuals who are reclassified
Adaptive training	• Apply to subset of skills that can be objectively assessed	• Apply to specialties that use proficiency-based readiness assessments (e.g., medical) • Apply to skills that can be rehearsed asynchronously
Assignment recommendations	• Use the Instrument Talent Marketplace to gather performance and satisfaction data • Focus on select workforce segments (e.g., by grade or specialty) • Develop rule-based system to capture assignment teams' logic	• Deliver recommendations rather than automating decisions • Include training about how to use recommendations given in Talent Marketplace
Promotion recommendations	• Focus on large developmental categories for which more data exist • Store digitized OPRs, order of merit, and board scores in data warehouse	• Deliver summaries and recommendations rather than automating decisions • Apply to lower-stakes board processes
Predicting separations	• Establish feeds from HRM systems to central data warehouse	• Identify downstream processes for which average separation rates are inadequate

The green shapes in Figure 5.1 show where we place the five use cases after reducing their technical and nontechnical complexity. These simplifications may also reduce the utility of the system, however. For example, although it would be less complex to apply an ML system to reclassification than to initial classification, only a fraction of enlisted personnel are reclassified. Likewise, although it would be less complex to introduce adaptive training to occupations that already use proficiency-based readiness assessments, many occupations do not. Finally, although it would be less complex to introduce ML systems that deliver assignment and promotion recommendations, fully automating decisions would maximize human capital savings. Given these types of trade-offs, the DAF must decide whether benefits of reducing complexity offset the drawbacks of reducing business value.

Summary

To balance the demands of multiple AI use cases, the DAF should consider the sources of technical and nontechnical complexity that they entail. Evaluating use cases in this way will allow the DAF to align the portfolio of AI use cases with the types of resources needed to implement them.

Across five use cases, we identified 20 different sources of technical and nontechnical complexity. Some, such as accessing data and upskilling the workforce, were pervasive. Others, such as dealing with measurement limitations, were less common.

Importantly, for all use cases, we identified variations in implementation intended to reduce complexity. Although these hedging actions increase the odds of success, they also reduce business value. At this stage of evaluation, the DAF may consider whether the trade-off is justified.

Chapter 6. Step 5. Align Project Portfolio with Available Analytic Resources

The final step of the framework uses the business value and complexity assessments of potential use cases to form a project portfolio. We recommend a portfolio rather than a prioritization approach because decisionmakers must make risk-versus-reward trade-offs when considering the different project dimensions. The goal of building an innovation portfolio is to balance lower-risk initiatives that may produce incremental gains with higher-risk initiatives that may be transformative.

Innovation Management

Figure 6.1 shows the ubiquitous effort-value matrix.[27] The matrix conceptualizes projects along two dimensions: business value and complexity. The intuition is that decisionmakers should select projects with high business value and low complexity (upper-left quadrant), and they should avoid ones with low business value and high complexity (lower-right quadrant). Decisionmakers should allocate remaining resources to a mix of low-risk but moderate-impact projects along with high-risk but potentially transformative ones (lower-left and upper-right quadrants). Thus, as compared with projects in the upper-left and lower-right quadrants, which permit go or no-go judgments, decisionmakers must engage in additional deliberation to decide among projects that trade off business value and complexity.

[27] Stackowiak and Kelly, 2020.

Figure 6.1. Notional Effort-Value Matrix for Decisionmaking

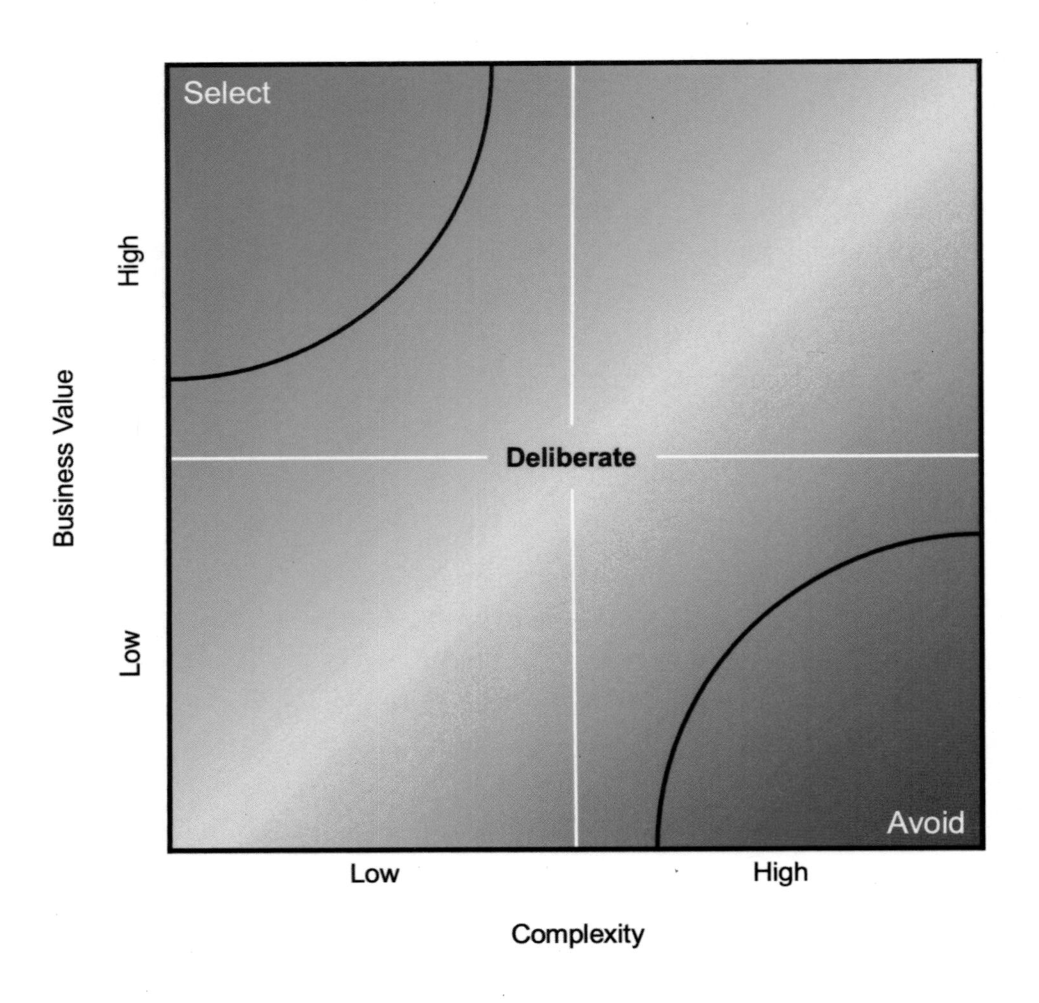

SOURCE: Adapted from Stackowiak and Kelly, 2020.

To excel at innovation management, organizations must balance investments at three levels of ambition (Figure 6.2).[28] Horizon 1 initiatives seek to make incremental improvements to established business processes. Whether it is connecting systems to automatically flag conflicts between planned deployments and medical appointments, using ML models to recommend promotions, or automatically recommending course schedules to reduce white space in training pipelines, these innovations draw on assets and processes that are already in place.

Horizon 2 initiatives accomplish existing business processes in new ways that are more efficient and effective. Whether it is using ML models to recommend occupations for service members, applying adaptive algorithms to tailor the training that service members receive, or

---

[28] Mehrdad Baghai, Stephen Coley, and David White, *The Alchemy of Growth: Practical Insights for Building the Enduring Enterprise*, Perseus Publishing, 2000.

using an RL agent to recommend changes to HYT, these innovations apply new technologies to perform established processes in new ways.

Horizon 3 initiatives create entirely new business processes that better meet the organization's needs.[29] Whether it is using ML and NLP to deliver career coaching to all service members regardless of rank and experience, using ML and recommendation engines to enable fully distributed decisionmaking in the Talent Marketplace, or using an RL agent to continuously adjust SRB and compensation levels by workforce segment to meet workforce needs, these innovations leverage technologies to conduct business in altogether new ways.

Figure 6.2. Three Horizons for Innovation Management

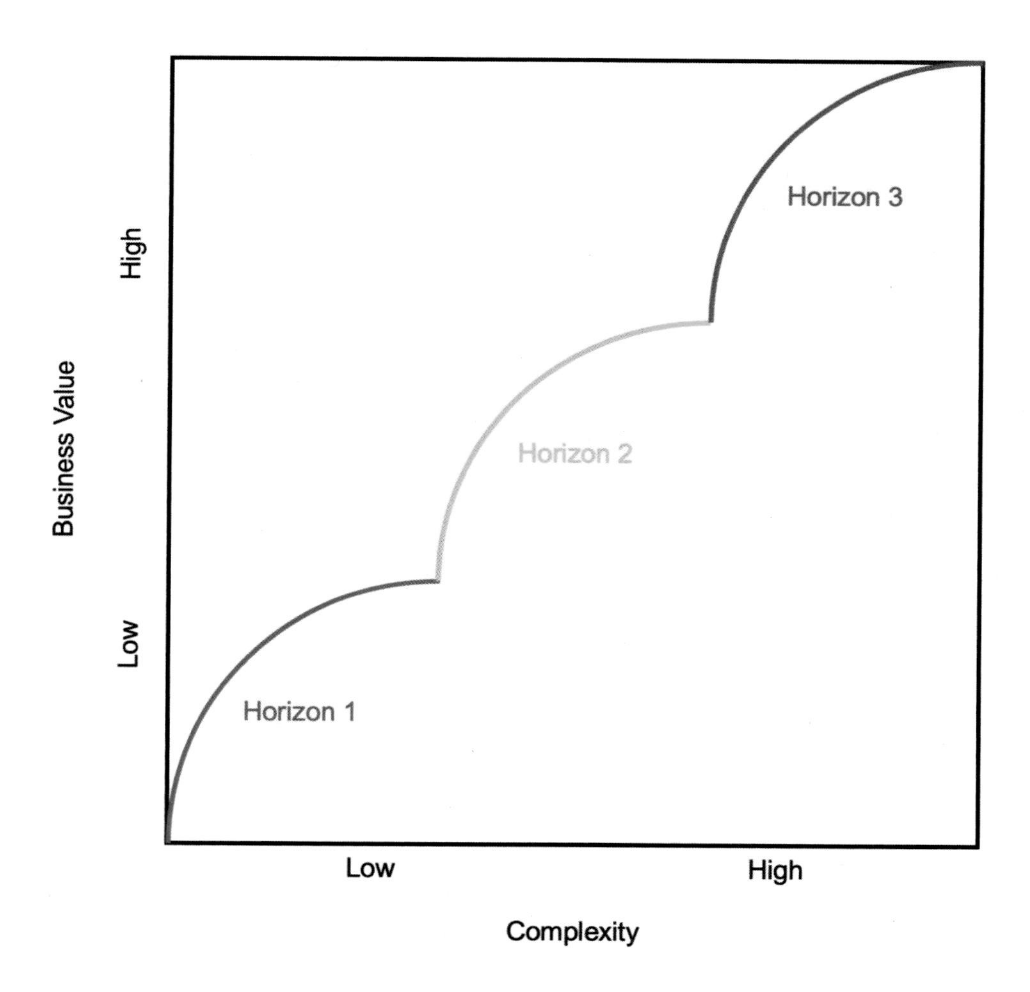

SOURCE: Adapted from Baghai, Coley, and White, 2000.

---

[29] Horizon 3 initiatives feature high complexity and business value. However, some projects with high complexity and business value do not involve creating entirely new business processes and thus would not meet the definition of a Horizon 3 initiative.

Research from private companies is informative with respect to the proper balance for an innovation portfolio. Across sectors, companies that allocate 70 percent of innovation activity to core business processes (i.e., Horizon 1 initiatives), 20 percent to adjacent ones (i.e., Horizon 2 initiatives), and 10 percent to transformative ones (i.e., Horizon 3 initiatives) outperformed their peers.[30] The 70-20-10 balance is not a golden ratio—it is simply a cross-industry average. The right balance depends on such factors as the pace of innovation in the sector, the company's competitive position, and the company's stage of development. Given that the pace of innovation in HRM is slow, and that the USAF is an established and competitive employer, relatively greater emphasis on strengthening core processes is a suitable balance.

Reducing Complexity

The bottom axis in Figure 6.1 combines technical and nontechnical complexity to simplify the visual. The former concerns technical barriers (e.g., algorithm development) that must be overcome for the innovation to succeed, whereas the latter concerns nontechnical barriers (e.g., organizational, culture, and policy changes). In practice, the two types of complexity must remain distinct because different strategies are needed to buy down technical versus nontechnical complexity (Table 6.1). For example, change management communications and business case analysis may increase buy-in from senior leaders and end-users. Modeling and simulation may establish potential benefits along with technical performance parameters. Additionally, advanced technology demonstrations and usability studies may increase technological readiness levels (TRLs) while allowing end-users to provide input and to shape concepts of operation. Finally, basic and applied research may mature the scientific principles to enable the technology.[31]

---

[30] Bansi Nagji and Geoff Tuff, "Managing Your Innovation Portfolio," *Harvard Business Review*, Vol. 90, No. 5, May 2012.

[31] Aside from requiring different organizations and strategies, reducing technical and nontechnical complexity may also depend on different types of funding.

Table 6.1. Strategies for Buying Down Complexity

Activity	Technical Complexity	Nontechnical Complexity
Change management communication		+
Workforce training		+
Business case analysis		+
Define metrics to demonstrate ROI		+
Phased implementation and rollout strategy	+	+
Experimental campaigns	+	+
Usability studies	+	+
System development and demonstration	+	+
Advanced technology demonstration	+	+
Modeling and simulation	+	+
Landscape and horizon analysis	+	
Applied research	+	
Basic research	+	

Contributions from different DAF and federal entities are also needed to reduce different forms of complexity (Table 6.2). For example, the Air Force Deputy Chief of Staff for Manpower, Personnel, and Services (AF/A1) and the Assistant Secretary of the Air Force, Manpower and Reserve Affairs (SAF/MR) can socialize concepts to increase high-level buy-in, and they can issue policy changes to take full advantage of affordances of AI systems. The Director of Plans and Integration can develop personnel plans to upskill the workforce and invest in IT systems needed to field AI. Additionally, AFWERX can conduct experimental campaigns to socialize concepts with end-users and to achieve higher TRLs. These might include, for example, using AI systems in operationally relevant environments with end-users to determine how well the systems work and how they can be best leveraged. Finally, academia and industry can develop new AI approaches applicable to HRM.

Table 6.2. Organizations for Buying Down Complexity

Entity	Technical Complexity	Nontechnical Complexity
Congress		+
AF/A1		+
Assistant Secretary of the Air Force, Manpower and Reserve Affairs (SAF/MR)		+
Major Commands		+
Director of Plans and Integration (AF/A1X)	+	+
Air Force Personnel Center (AFPC)	+	+
Federally funded research and development centers	+	+
Defense industry base	+	+
AFWERX	+	+
Department of the Air Force, Chief Data Office	+	
The Air Force Research Laboratory	+	
Joint Artificial Intelligence Center	+	
Private-sector research and development	+	
Academic research and development	+	

Application

Using the effort-value matrix (Figure 6.1), the USAF could select a portfolio of AI projects to meet different objectives. A study of companies in the industrial, technology, and consumer goods sectors identified three general innovation portfolio investment patterns.[32] Leading companies in stable sectors predominantly invested in Horizon 1 initiatives to yield near-term gains. Leading companies in dynamic sectors predominantly invest in Horizon 1 and 2 initiatives to yield near-term gains while allowing themselves to maintain their position as the sector changes. Finally, lagging companies or ones in emerging sectors invest more in Horizon 2 and 3 initiatives to establish their position in the first place.

If the goal is to maximize near-term gains, the USAF should focus on projects with high business value and low complexity (i.e., the upper-left quadrant of Figure 6.1). Even among projects contained in that quadrant, the USAF could build a portfolio that spans phases of the HRM life cycle, HRM business objectives, or core ML approaches (Table 6.3).

If the goal is to secure near-term gains while creating midterm opportunities, the USAF should focus on projects with high business value (i.e., both upper quadrants of Figure 6.1). Once again, the USAF could build a portfolio that spans phases of the HRM life cycle, HRM business objectives, or core ML approaches. However, an additional consideration for projects in the

[32] Nagji and Tuff, 2012.

upper-right quadrant of Figure 6.1 is the balance between technical and nontechnical complexity, and the types of resources needed for each project.

Finally, if the goal is to advance the AI horizon and, in so doing, to create significant mid- to long-term opportunities, the USAF should undertake some projects with high complexity and high business value (i.e., the upper-right quadrant of Figure 6.1). Although these studies lack immediate business value, they may overcome feasibility challenges and thereby enable future projects with significant business value (Table 6.3).

Table 6.3. Example Portfolios

Objective	Projects
Maximize near-term gains	Accession date–to-course scheduling algorithm (Horizon 1) Occupational classification (Horizon 2) Resource allocation to decrease SAT status (Horizon 1) Set SRB levels (Horizon 2)
Secure near-term gains while creating midterm opportunities	Accession date-to-course scheduling algorithm (Horizon 1) Resource allocation to decrease SAT status (Horizon 1) Occupational classification (Horizon 2) Give assignment recommendations (Horizon 3)
Create significant mid- to long-term opportunities	Adaptive training (Horizon 3) Career coaching (Horizon 3) Compensation planning (Horizon 3) DE recommendations (Horizon 2)

Summary

As the USAF considers projects that apply AI to HRM, it must balance investments across different levels of innovation ambition (i.e., the three horizons). The effort-value matrix is a useful construct for arranging initiatives by complexity and business value and for building a portfolio of projects to meet the USAF's immediate and future needs.

High-risk, high-reward initiatives are an essential part of an innovation portfolio. Yet these initiatives present unique challenges in terms of technical and implementation complexity. As the USAF chooses among high-risk, big-reward initiatives, it must consider the organizations, strategies, and types of resources needed to buy down these different forms of complexity.

Chapter 7. Conclusions and Recommendations

DAF processes for prioritizing analytic projects already reflect the wisdom of using advisory panels of stakeholders with objective criteria to reach a consensus about which projects have the best chance of helping the broader organization to fulfill its mission. Our exploration of the challenge of evaluating and vetting projects seeking to develop AI decision-support systems has certainly not revealed the need to depart from this tried-and-true approach. However, this research does suggest the following conclusions, each with recommendations for the DAF to consider as it follows the DoD direction to become a "data-centric" organization.

Notably, our framework presupposes a particular solution class—AI systems for HRM. Thus, it is not intended to replace strategic resource management processes. Nevertheless, the framework is useful for organizations not resourced to consider all possible technical and nontechnical solutions for a particular HRM problem, and it is useful for rejecting AI projects that are unlikely to narrow significant capability gaps.

Conclusion 1. The proliferation of DAF data platforms and analytic tools have created the potential for a wide range of applications of AI to HRM.

AI can be used to net efficiency gains and cost savings from existing DAF HRM processes. In addition, it can be used to enable new HRM processes that more fully leverage DAF human capital to create competitive advantage. The 19 use cases in Chapter 2 illustrate just some of the ways that AI can be used in the HRM domain.

Recommendation 1a. To maximize ROI, the DAF must use a systematic process to evaluate AI projects for HRM and to build a balanced portfolio.

The decision process begins by linking use cases to the HRM objectives they satisfy. Next, the process identifies and evaluates the technical feasibility of implementing AI solutions. The process then considers the complexity of fielding the solutions. Although this approach does not guarantee that projects will succeed, it places investments where they are most likely to yield value and insight. This function could be performed by the Deputy Chief for Manpower, Personnel and Services, Headquarters U.S. Air Force and would require inputs from the Office of the Secretary of the Air Force and from major commands as well. Any organization building out a portfolio of AI projects for HRM could also use features of this framework. Ideally, the panel of individuals performing the process will include some with domain knowledge of DAF HRM processes and others with technical knowledge of AI.

Conclusion 2. A common set of ML approaches was applicable to most of the use cases.

Of the common ML approaches under development—supervised learning, optimization, NLP, unsupervised learning, and RL—all played a role in multiple potential use cases. These classes of methods have been extensively researched and developed, and they have been increasingly used for real-world problems including HRM.

Recommendation 2a. The DAF should develop a common ML ecosystem to enable rapid creation, comparison, and reuse of ML pipelines, models, and DoD datasets.

The underlying structure for many of the prediction and decision problems we considered was similar. Thus, methods and models that work for one problem are likely to work for others. A common ecosystem would standardize workflows, and it would enable reuse of ML capabilities across HRM problems. Envision, a new readiness-tracking tool, provides a viable platform to host such an ecosystem.

Recommendation 2b. DoD should maintain an innovation dashboard of ML projects for HRM.

The complete portfolio of AI projects for HRM is spread across the DAF, and no one office has visibility into all initiatives. Maintaining an innovation dashboard would allow the DAF to better manage innovation across the broader portfolio of AI projects for HRM. Better management could include ensuring that projects are not reproposed once they are deemed infeasible until strategic initiatives can address the feasibility barriers, as well as holding organizations accountable for whether they implement AI initiatives and realize the predicted ROI (thus bridging the "valley of death"). An innovation dashboard could also facilitate the reuse of ML capabilities across HRM problems.

Conclusion 3. Availability of suitable data, rather than availability of suitable methodologies, was the greatest technical barrier.

For most use cases we considered, a suitable AI methodology exists. However, for many use cases, suitable inputs, measurable outputs, or both do not currently exist. This limited the feasibility of applying AI for well over half of the HRM problems we considered.

Recommendation 3a. To enable applications of AI to HRM, the DAF must continue to invest in data infrastructure and outcome definitions.

The DAF must identify factors that could conceivably contribute to the outcome of interest, explore methods to capture those variables, and make them available for ML modeling efforts. In addition to expanding the space of input variables, the DAF must settle on context-specific definitions of outcomes such as *training success*, *career success*, and *satisfaction*. As part of this recommendation, the DAF should coordinate with domain experts to ensure that the right measures are selected.

Conclusion 4. An innovation portfolio includes low-risk/low-reward projects along with higher-risk but potentially transformative ones.

The conventional wisdom from private-sector companies is that about 70 percent of a research portfolio should be allocated to core business functions, and the remaining 30 percent should be allocated to more-transformative initiatives. This yields near-term wins while maturing more-innovative concepts.

Recommendation 4a. The DAF should begin by prioritizing lower-risk AI projects in the near term, given the potentially significant value of many of those projects.

This conventional wisdom applies to the DAF. Given the underutilization of AI for HRM, many low-risk/high-reward opportunities exist. The DAF should focus on these to first establish the utility of AI for HRM. According to our assessment of HRM objectives, many AI systems enable measurable process improvements. Although the DAF should consider other HRM objectives as well, AI systems with quantifiable ROI may be particularly compelling from the onset.

Recommendation 4b. The DAF should allocate a smaller percentage of resources to higher-risk projects.

Even as the DAF solidifies near-term gains, if resources permit, it should undertake some projects with significant feasibility or complexity concerns. By removing these barriers, the DAF can create significant sources of future business value.

Conclusion 5. Implementation may be complicated by technical and nontechnical complexity.

We noted 20 different sources of technical and nontechnical complexity across use cases. These do not eliminate the possibility of a project succeeding; however, they do call for foresight and planning.

Recommendation 5a. As the DAF evaluates projects, it must consider the types of complexity they entail.

Different types of capital and different strategies are needed to buy down these different forms of complexity. As the DAF evaluates projects, it must align resources to address the complexities they entail. Additionally, the DAF must consider whether, for a particular use case, there are system design options that present a more attractive balance between complexity and business value.

Appendix. Use Cases for Complexity Analysis

In this appendix, we describe the sources of technical and nontechnical complexity for the five use cases we considered. In order of phases of the HRM life cycle, these are occupational classification, adaptive training, assignment recommendations, promotion recommendations, and predicting separations.

Occupational Classification

Most enlisted personnel enter BMT with a Guaranteed Training Enlistment Program (GTEP) contract. However, some are classified (i.e., assigned to a specialty) during BMT, and some are reclassified after they fail to complete initial skills training. The consequences of career field classification are significant in terms of training success, job performance, and career outcomes.[33]

An AI system could improve occupational classifications by learning from historical data the attributes of airmen associated with training and career success in different specialties. Once trained, these models can be used to predict which specialties new airmen are most likely to succeed in. Additionally, given that there are a fixed number of training seats available for each specialty, the AI system can consider how to assign all individuals to achieve the best overall outcomes.

Technical Complexity

This use case entails numerous sources of technical complexity. To predict training and career outcomes, data must be accessed from multiple different administrative systems and combined. New data sources, such as the Air Force Work Interest Navigator (AF-WIN), that may improve classification are available only for a limited number of years and airmen. Additionally, historical data sources may contain measures from extant tests. Finally, some inputs, such as composite scores from the Armed Services Vocational Aptitude Battery (ASVAB), are used to screen individuals, making it difficult to predict outcomes for those who fall below cutoffs.[34]

Occupational classification changes over time. Information on basic eligibility and decisions affecting entry into career fields has not always been retained, making it difficult to account for

---

[33] Sean Robson, Maria C. Lytell, Matthew Walsh, Kimberly C. Hall, Kirsten M. Keller, Vikram Kilambi, Joshua Snoke, Jonathan W. Welburn, Patrick S. Roberts, Owen Hall, and Louis T. Mariano, *U.S. Air Force Enlisted Classification and Reclassification: Potential Improvements Using Machine Learning and Optimization Models*, RAND Corporation, RR-A284-1, 2022.

[34] In the psychometric literature, this is called *range restriction*.

effects of historical policies. In addition, some specialties have been eliminated or combined, and other, new specialties have been created. This obviates some historical data and introduces the challenge of predicting outcomes for entirely new specialties.

Finally, outcomes such as job performance and career success can be hard to measure and equate across jobs; they are affected by myriad factors and span far into the future. Aside from making it more difficult to train an ML model, these factors make it more difficult to test the model in an operational environment and to demonstrate its utility.

Nontechnical Complexity

This use case also entails numerous sources of nontechnical complexity. Occupational classification has multiple stakeholders, including individual airmen, Air Education and Training Command (AETC), AF/A1, the Surgeon General, Air Force Recruiting Service, and career field managers. The DAF must establish buy-in across these stakeholder groups. Relatedly, stakeholders may disagree about the relative importance of personal preferences, training success, job performance, and career outcomes. Finally, frontline workers must trust the AI system, as must the individuals whom decisions affect.

This use case also requires changes to existing processes. Currently, almost 90 percent of contracts are GTEP. In addition, when enlisted personnel are reclassified, the primary determinant is minimizing wait time. More-flexible processes for initial classification and reclassification are needed to maximize the utility of the AI system.

Lastly, this use case, along with all other use cases, requires a workforce with technical skills to build, train, evaluate, and deploy the AI system.

Adaptive Training

Most DAF training and education occurs in lockstep. Individuals advance through a fixed curriculum and, in the case of synchronous learning, at a fixed pace. There are limited opportunities to tailor the timing and content of training to the individual.

A one-size-fits-all approach does not optimize training resources. Some individuals receive more training than they need, or they complete training events that do not address their greatest weaknesses. Worse yet, this approach creates risk when proficiency is assumed but not truly measured after individuals complete a prescribed curriculum.

AI can be used to deliver personalized training.[35] For example, AI methods can trace an individual's level of mastery of KSAs, along with skill decay that occurs during periods of

---

[35] Tiffany S. Jastrzembski, Matthew Walsh, Michael Krusmark, Suzan Kardong-Edgren, Marilyn Oermann, Karey Dufour, Teresa Millwater, Kevin A. Gluck, Glenn Gunzelmann, Jack Harris, and Dimitrios Stefanidis, "Personalizing Training to Acquire and Sustain Competence Through Use of a Cognitive Model," in D. Schmorrow and C. Fidopiastis, eds., *Augmented Cognition: Enhancing Cognition and Behavior in Complex Human Environments*, conference paper, Springer, 2017.

disuse. This information can be used to determine which training events to deliver and when. Adaptive training can be used in a variety of contexts, ranging from initial skills training to pilot continuation training.

Technical Complexity

To conduct adaptive training, the KSAs underlying task performance must be identified. KSAs must also be linked to learning experiences and to objective performance measures that can be automatically assessed. This creates complexity in terms of authoring and tagging training content and assessment materials, instrumenting the environment to capture performance data, and creating algorithms to score performance. Further, because adaptive training builds on an individual's past experiences, it requires a learning management system to store detailed information about the individual's performance history.

Nontechnical Complexity

Adaptive training requires a fundamental shift in thinking about readiness from frequency-based to performance-based.[36] This may create friction, both in terms of the judgment attached to individuals who require more training to meet new proficiency definitions, and the risk of giving certain (high-performing) individuals less training.

Current processes are also not designed to support flexible training pathways. For example, students advance through initial skills training pipelines in lockstep. Adaptive training would be most beneficial if high-performing students could advance through the training pipeline faster than their peers.[37] Likewise, pilot upgrade and continuation training are based on predictable event counts prescribed by squadron training syllabi and by the Ready Aircrew Program. As the saying goes, "plan what you want, fly what you can, log what you need."[38] Adaptive training would be most beneficial if squadrons had resources and flexibility to support the true but less predictable needs of their aircrews.

Lastly, this use case would place greater demands on instructors and educational designers to create learning experiences and assessments and to link them to underlying KSAs.

Assignment Recommendations

The USAF recently adopted a Talent Marketplace. Officers eligible for assignment may post a profile, view available positions, and submit a list of position preferences to the Talent

---

[36] Robert Chapman and Charles Colegrove, "Transforming Operational Training in the Combat Air Forces," *Military Psychology*, Vol. 25, No. 3, 2013.

[37] AETC has begun to use proficiency advancement in narrow cases.

[38] Bill Taylor, former PAF senior mathematician and Air Force readiness expert, personal communication with authors, August 3, 2022.

Marketplace. Likewise, position owners may post vacancies, review officer profiles, and submit a list of preferences for officers included in the assignment cycle. Assignment teams review preferences and, after taking other considerations into account, make assignment decisions to balance the needs of the Air Force along with individuals' and position owners' preferences.

An AI system could enhance assignments by recommending positions to officers that would maximize their development and satisfaction. Likewise, the system could recommend candidates to position holders. Finally, the system could apply an algorithm to match officer and position holder preferences to arrive at a preliminary solution for assignment teams to refine.[39]

Technical Complexity

As with occupational classification, generating assignment recommendations requires accessing data from multiple administrative systems. In addition, some sources of data, such as position descriptions, are expressed using natural language. NLP methods are needed to convert these semistructured inputs to a form suitable for ML models. Further, assignment outcomes can be evaluated across multiple, often competing dimensions, such as personal satisfaction, performance, development, and relationship to promotion outcomes. Some of these outcomes, such as satisfaction and performance, are hard to measure.

Aside from the challenges of gathering inputs and defining outputs, the size of the decision space is large. Hundreds to thousands of officers and positions must be matched within a given assignment cycle. The AI system should consider which officers and positions are interchangeable.

The assignment problem changes over time. Historical assignments reflect policies that may no longer be in effect. The AI system should not learn to perpetuate these historical patterns. In addition, some positions have been eliminated or their duties have changed, rendering historical data obsolete, and other, new positions have been created, going beyond what can be inferred from historical data.

Finally, although a large amount of data are available to train an ML model, the number of development pathways within and between specialties is vast. For example, officers in different specialties may be treated differently, as may be individuals with high versus moderate potential. It may be difficult to train an ML model to learn effective development recommendations given the available data.

Nontechnical Complexity

Assignments affect all officers, commanders, and functional communities. Thus, implementing an AI system for assignment recommendations would require widespread buy-in. Complicating matters, different stakeholders have different objectives. For example, officers

[39] This topic is explored further in Calkins et al., 2024.

may seek to maximize satisfaction, position owners may seek to maximize current performance, and assignment teams may seek to maximize future performance. Creating a system that balances these objectives may be difficult.

If officers, position holders, and assignment teams are to use the AI system, they must receive training and education, and they must come to trust the system.

Promotion Recommendations

Each year, the DAF invests hundreds of staff hours, drawn from some of the most senior military leaders, to convene promotion boards to decide which officers have the greatest potential to serve at higher ranks. Aside from being resource-intensive, the outcomes of promotion boards are vital to the health of the future force.

An AI system could replace or augment board processes, including promotion boards. An AI model could be trained using OPRs and promotion outcomes from historical boards. When given new officer records, the system could use the trained model to make decisions, to give recommendations to human raters, to summarize records for human raters, to deliver feedback to officers being considered, or to cross-check human board decisions for fairness and consistency.[40]

Technical Complexity

As with occupation classification and assignments, the inputs needed to predict promotion outcomes are contained across multiple different administrative systems. In addition, OPRs, a major input to promotion decisions, are expressed as semistructured narrative text. NLP methods are needed to prepare these inputs for ML models. Complicating matters, older OPRs, though digitized, are not machine readable.

Aside from challenges with inputs, the outcome (i.e., promote or do not promote) is impoverished. Selection rates are near 100 percent for some ranks. Order of merit or board scores, if available, would reduce the technical challenge of training ML models to predict outcomes with so little variability.

Promotion policies have also changed, with the Line of the Air Force being divided into separate developmental categories, and the DAF doing away with below-the-zone promotions. An indirect effect of these changes is that historical promotion data may be less useful for training models to predict future outcomes.

A final, fundamental technical barrier is that the ground truth is unknown. In a sense, the goal for the ML model is to replicate historical selections. However, historical selections are susceptible to human error and bias. Thus, historical selections are a proxy for the variable of

---

[40] See David Schulker, Joshua Williams, Cheryl K. Montemayor, Li Ang Zhang, and Matthew Walsh, *The Personnel Records Scoring System: A Methodology for Designing Tools to Support Air Force Human Resources Decisionmaking*, RAND Corporation, RR-A1745-3, 2024.

interest—fitness of an officer to serve at higher ranks. In cases where the ML model disagrees with human raters, it may be unclear who is correct.

Nontechnical Complexity

Promotions are high-stakes decisions for individuals and for the DAF, and the promotion process is highly visible. As a result, any attempt to change officer promotions will undergo significant scrutiny. In addition, any changes to officer promotions will require buy-in from senior leaders, board members, and officers affected by outcomes.

In some implementation designs, the outputs of an AI model may be given to human board members in the form of recommendations or summaries. An additional source of nontechnical complexity in these cases is delivering training and education to allow board members to properly use system outputs.

Predicting Future Separation

Managing employee retention is important to any organization. This is especially true for the DAF, which must hire, develop, and retain individuals for technical and management roles over careers spanning multiple decades. Aside from detecting negative retention trends, retention forecasts are needed for multiple HRM functions, such as accession and compensation planning.

The most prevalent approach to forecasting retention is to calculate average separation rates over selected previous periods, usually within demographic groups defined by specialty and years of service. An AI system could be trained to generate more-precise estimates using information about service member characteristics, attitudes and perceptions, and environmental characteristics.[41] Model predictions could be used to construct more-accurate retention forecasts for other, downstream processes.

Technical Complexity

In a sense, technical complexity is low. This is a standard supervised learning problem with a well-defined outcome (i.e., separation) and a large amount of historical data. However, as with earlier examples, the data needed to predict separations are contained across multiple different administrative systems. In addition, the outcome of interest—monthly separations—occurs with very low frequency, complicating how ML models may be trained. Moreover, important inputs, such as service member attitudes and perceptions, are not routinely gathered. Finally, historical loss rates may reflect the influences of policies, such as force separation or stop loss, that are no longer in effect.

---

[41] David Schulker, Lisa M. Harrington, Matthew Walsh, Sandra K. Evans, Irineo Cabreros, Dana Udwin, Anthony Lawrence, Christopher E. Maerzluft, and Claude M. Setodji, *Developing an Air Force Retention Early Warning System: Concept and Initial Prototype*, RAND Corporation, RR-A545-1, 2021.

Nontechnical Complexity

The current method for generating retention forecasts (e.g., three-year averages) is reasonably accurate and readily interpretable. Thus, one source of nontechnical complexity is replacing a viable alternative. Another potential source of nontechnical complexity is that downstream processes have come to rely on stable, albeit imperfect, outputs from current processes. An ML model may give more accurate forecasts, yet it may be difficult to incorporate highly variable projections into planning processes.

Abbreviations

AETC	Air Education and Training Command
AF/A1	Deputy Chief of Staff for Manpower, Personnel, and Services
AI	artificial intelligence
BMT	basic military training
DAF	Department of the Air Force
DE	developmental education
DoD	U.S. Department of Defense
DT	development team
GTEP	Guaranteed Training Enlistment Program
HRM	human resource management
HYT	high year tenure
KSAs	knowledge, skills, and abilities
ML	machine learning
MWR	morale, welfare, and recreation
NLP	natural language processing
OPR	officer performance report
RL	reinforcement learning
ROI	return on investment
SAT	students awaiting training
SMART	specific, measurable, achievable, relevant and time-bound
SRB	selective reenlistment bonus
TRL	technological readiness level
USAF	U.S. Air Force
VAULTIS	visible, accessible, understandable, linked, trustworthy, interoperable, and secure
WAPS	Weighted Airman Promotion System

References

Baghai, Mehrdad, Stephen Coley, and David White, *The Alchemy of Growth: Practical Insights for Building the Enduring Enterprise*, Perseus Publishing, 2000.

Budhwar, Pawan, Ashish Malik, M. T. Thedushika De Silva, and Praveena Thevisuthan, "Artificial Intelligence–Challenges and Opportunities for International HRM: A Review and Research Agenda," *International Journal of Human Resource Management*, Vol. 33, No. 6, 2022.

Calkins, Avery, Monique Graham, Claude Messan Setodji, David Schulker, and Matthew Walsh, *Machine Learning–Enabled Recommendations for the Air Force Officer Assignment System:* Vol. 5, RAND Corporation, RR-A1745-5, 2024.

Chapman, Pete, Julian Clinton, Randy Kerber, Thomas Khabaza, Thomas Reinartz, Colin Shearer, and Rüdiger Wirth, *CRISP-DM 1.0: Step-by-Step Data Mining Guide*, NCR Systems Engineering Copenhagen, DaimlerChrysler, SPSS Inc., and OHRA Verzekeringen en Bank Groep, 2000.

Chapman, Robert, and Charles Colegrove, "Transforming Operational Training in the Combat Air Forces," *Military Psychology*, Vol. 25, No. 3, 2013.

Daneva, Maya, Egbert van der Veen, Chintan Amrit, Smita Ghaisas, Klaas Sikkel, Ramesh Kumar, Nirav Ajmeri, Uday Ramteerthkar, and Roel Wieringa, "Agile Requirements Prioritization in Large-Scale Outsourced System Projects: An Empirical Study," *Journal of Systems and Software*, Vol. 86, No. 5, May 2013.

DoD—*See* U.S. Department of Defense.

Emmerichs, Robert M., Cheryl Y. Marcum, and Albert A. Robbert, *An Operational Process for Workforce Planning*, RAND Corporation, MR-1684/1-OSD, 2004. As of July 12, 2023: https://www.rand.org/pubs/monograph_reports/MR1684z1.htm

Englund, Randall L., and Robert J. Graham, *Creating an Environment for Successful Projects*, 3rd ed., Berrett-Koehler Publishers, 2019.

Fountaine, Tim, Brian McCarthy, and Tamim Saleh, "Building the AI-Powered Organization," *Harvard Business Review*, July–August 2019.

Guenole, Nigel, and Sheri Feinzig, *The Business Case for AI in HR, with Insights and Tips on Getting Started*, IBM Smarter Workforce Institute, IBM Corporation, 2018.

Guenole, Nigel, Jonathan Ferrar, and Sheri Feinzig, *The Power of People: Learn How Successful Organizations Use Workforce Analytics to Improve Business Performance*, Pearson Education, 2017.

Gutiérrez, E., I. Kihlander, and J. Eriksson, "What's a Good Idea? Understanding Evaluation and Selection of New Product Ideas," in *DS 58-3: Proceedings of ICED 09, the 17th International Conference on Engineering Design*, Vol. 3, *Design Organization and Management*, Design Society, 2009.

Jastrzembski, Tiffany S., Matthew Walsh, Michael Krusmark, Suzan Kardong-Edgren, Marilyn Oermann, Karey Dufour, Teresa Millwater, Kevin A. Gluck, Glenn Gunzelmann, Jack Harris, and Dimitrios Stefanidis, "Personalizing Training to Acquire and Sustain Competence Through Use of a Cognitive Model," in D. Schmorrow and C. Fidopiastis, eds., *Augmented Cognition: Enhancing Cognition and Behavior in Complex Human Environments*, conference paper, Springer, 2017.

Jiang, Kaifeng, David P. Lepak, Jia Hu, and Judith C. Baer, "How Does Human Resource Management Influence Organizational Outcomes? A Meta-Analytic Investigation of Mediating Mechanisms," *Academy of Management Journal,* Vol. 55, No. 6, 2012.

Knapp, David, Bruce R. Orvis, Christopher E. Maerzluft, and Tiffany Berglund, *Resources Required to Meet the U.S. Army's Enlisted Recruiting Requirements Under Alternative Recruiting Goals, Conditions, and Eligibility Policies*, RAND Corporation, RR-2364-A, 2018. As of July 12, 2023:
https://www.rand.org/pubs/research_reports/RR2364.html

Krumpal, Ivar, "Determinants of Social Desirability Bias in Sensitive Surveys: A Literature Review," *Quality and Quantity*, Vol. 47, No. 4, 2013.

Meadows, Sarah O., Stephanie B. Holliday, Wing Y. Chan, Stephani L. Wrabel, Margaret Tankard, Dana Schultz, Christopher M. Busque, Felix Knutson, Leslie A. Payne, and Laura L. Miller, *Air Force Morale, Welfare, and Recreation Programs and Services: Contribution to Airman and Family Resilience and Readiness—Appendixes*, RAND Corporation, RR-2670-AF, 2019. As of July 12, 2023:
https://www.rand.org/pubs/research_reports/RR2670.html

Myatt, Summer, "Air Force's Data Fabric in Maturation Stage, Officials Say," *GovConWire*, March 30, 2022. As of July 12, 2023:
https://www.govconwire.com/2022/03/air-forces-data-fabric-in-maturation-stage-officials-say/

Nagji, Bansi, and Geoff Tuff, "Managing Your Innovation Portfolio," *Harvard Business Review*, Vol. 90, No. 5, May 2012.

National Academies of Sciences, Engineering, and Medicine, *Strengthening U.S. Air Force Human Capital Management: A Flight Plan for 2020–2030*, National Academies Press, 2020.

Paul, Christopher, Jessica Yeats, Colin P. Clarke, Miriam Matthews, and Lauren Skrabala, *Assessing and Evaluating Department of Defense Efforts to Inform, Influence, and Persuade: Handbook for Practitioners*, RAND Corporation, RR-809/2-OSD, 2015. As of July 12, 2023: https://www.rand.org/pubs/research_reports/RR809z2.html

Pope, Adrian P., Jaime S. Ide, Daria Mićović, Henry Diaz, David Rosenbluth, Lee Ritholtz, Jason C. Twedt, Thayne T. Walker, Kevin Alcedo, and Daniel Javorsek, "Hierarchical Reinforcement Learning for Air-to-Air Combat, in *2021 International Conference on Unmanned Aircraft Systems (ICUAS)*, Institute of Electrical and Electronics Engineers, 2021.

Ransbotham, Sam, David Kiron, Philipp Gerbert, and Martin Reeves, "Reshaping Business with Artificial Intelligence: Closing the Gap Between Ambition and Action," *MIT Sloan Management Review*, Massachusetts Institute of Technology, 2017.

Robson, Sean, Maria C. Lytell, Matthew Walsh, Kimberly C. Hall, Kirsten M. Keller, Vikram Kilambi, Joshua Snoke, Jonathan W. Welburn, Patrick S. Roberts, Owen Hall, and Louis T. Mariano, *U.S. Air Force Enlisted Classification and Reclassification: Potential Improvements Using Machine Learning and Optimization Models*, RAND Corporation, RR-A284-1, 2022. As of July 12, 2023: https://www.rand.org/pubs/research_reports/RRA284-1.html

Schulker, David, Lisa M. Harrington, Matthew Walsh, Sandra K. Evans, Irineo Cabreros, Dana Udwin, Anthony Lawrence, Christopher E. Maerzluft, and Claude M. Setodji, *Developing an Air Force Retention Early Warning System: Concept and Initial Prototype*, RAND Corporation, RR-A545-1, 2021. As of July 12, 2023: https://www.rand.org/pubs/research_reports/RRA545-1.html

Schulker, David, Matthew Walsh, Avery Calkins, Monique Graham, Cheryl K. Montemayor, Albert A. Robbert, Sean Robson, Claude Messan Setodji, Joshua Snoke, Joshua Williams, and Li Ang Zhang, *Leveraging Machine Learning to Improve Human Resource Management:* Vol. 1, *Key Findings and Recommendations for Policymakers*, RAND Corporation, RR-A1745-1, 2024.

Schulker, David, Joshua Williams, Cheryl K. Montemayor, Li Ang Zhang, and Matthew Walsh, *The Personnel Records Scoring System:* Vol. 3, *A Methodology for Designing Tools to Support Air Force Human Resources Decisionmaking*, RAND Corporation, RR-A1745-3, 2024.

Secretary of the Air Force Public Affairs, "Chief Data Office Announces Capabilities for the VAULT Data Platform," October 11, 2019. As of July 12, 2023: https://www.af.mil/News/Article-Display/Article/1987254/chief-data-office-announces-capabilities-for-the-vault-data-platform/

Snoke, Joshua, Matthew Walsh, Joshua Williams, and David Schulker, *Safe Use of Machine Learning for Air Force Human Resource Management:* Vol. 4, *Evaluation Framework and Use Cases*, RAND Corporation, RR-A1745-4, 2024.

Society for Human Resource Management, "HR Glossary," webpage, undated. As of July 12, 2023: https://shrm.org/ResourcesAndTools/tools-and-samples/HR-Glossary/Pages/default.aspx

Stackowiak, Robert, and Tracey Kelly, *Design Thinking in Software and AI Projects: Proving Ideas Through Rapid Prototyping*, Springer Science and Business Media, 2020.

U.S. Department of Defense, *DoD Data Strategy*, September 30, 2020.

Walch, Kathleen, "How the Department of the Air Force Is Driving Forward with AI," *Forbes*, November 14, 2020. As of July 12, 2023: https://www.forbes.com/sites/cognitiveworld/2020/11/14/how-the-department-of-the-air-force-is-driving-forward-with-ai/?sh=15765900792c

Wright, Amy, Diane Gherson, Josh Bersin, and Janet Mertens, "Accelerating the Journey to HR 3.0: Ten Ways to Transform in a Time of Upheaval," IBM Institute for Business Value, IBM Corporation, 2020.